# Heat Advisory

## Protecting Health on a Warming Planet

Alan H. Lockwood, M.D.

T0296412

The MIT Press
Cambridge, Massachusetts
London, England

First MIT Press paperback edition, 2017
© 2016 Massachusetts Institute of Technology

This book was set in Sabon by Toppan Best-set Premedia Limited.

Library of Congress Cataloging-in-Publication Data

Names: Lockwood, Alan H., author.
Title: Heat advisory : protecting health on a warming planet / Alan H. Lockwood, M.D.
Description: Cambridge, MA : The MIT Press, [2016] | Includes bibliographical references and index.
Identifiers: LCCN 2016004065 | ISBN 9780262034876 (hardcover : alk. paper), ISBN 9780262534482 (paperback)
Subjects: LCSH: Global warming—Health aspects. | Climatic changes—Health aspects. | Medical anthropology.
Classification: LCC RA793 .L66 2016 | DDC 613/.1—dc23 LC record available at https://lccn.loc.gov/2016004065

To Anne—
and my family, present and future

# Contents

# Preface and Acknowledgments

The impetus to write this book arose as I was finishing *The Silent Epidemic: Coal and the Hidden Threat to Health*. As I wrote, it became clear that there was another, more important story that was larger than one limited to the consequences of burning coal—namely, the effects of climate change on health. Coal continues to exert important ill effects on health during all phases of its life cycle, particularly in countries such as China and India, but the tide is turning. All over the world, ordinary citizens, their governments, and in some cases the power companies themselves all seem to realize that burning coal is not an option if we are to move toward a more sustainable energy future. The once-limiting costs of wind and solar power have fallen precipitously even as some utilities seek to impose restrictive net metering fees on rooftop solar installations.

Utilities that take advantage of the solar option face new challenges. What will they do at night when the sun does not shine? Will a mix of improved battery technology and heat storage technologies fill this void? Or will research and development transform laboratory curiosities—such as the use of light to split water into easily stored hydrogen and oxygen— mature into commercially viable technologies? Wind turbines continue to improve, and plans are emerging to harness the energy of tides, waves, and rivers without building dams.

Chai Jing, a courageous and outspoken Chinese journalist and mother, produced a documentary called *Under the Dome* that she shows to huge audiences. Its impact was great enough that access to the documentary on YouTube was blocked in China. Now, China has entered into an agreement with President Obama to limit carbon dioxide emissions and has taken steps to initiate a cap and trade policy to control these emissions. Meanwhile, the *New York Times* reported data from the US Energy

Information Administration that shows that the BRIC countries (Brazil, the Russian Federation, India, and China) all generate a larger fraction of their electricity from renewable sources than we do in the United States. Of course, we Americans use much more electricity per capita even as we generate decreasing amounts from coal.

With this information in mind, I approached Clay Morgan, my editor at the MIT Press, about a follow-up to *The Silent Epidemic*. He and his associates were enthusiastic, and we agreed on a timeline that would see the book in the hands of readers before the 2016 presidential elections. Clay, like me, is now enjoying his retirement, and I am in the capable hands of Miranda Martin, Beth Clevenger, Kathleen Caruso, and their colleagues at the MIT Press. I am particularly indebted to the anonymous peer reviewers whose trenchant comments helped make this a better book. I owe a special debt of gratitude to Melinda Rankin whose invisible copyediting skills improved the style and accuracy of the final text.

The task of writing this book was daunting. Unlike those who made climate science their life's work, I became a clinical neuroscientist. In many ways, the research challenges I faced prepared me for the task I have undertaken. Success depended on reading trusted sources as widely as possible, making evidence-based decisions, resorting to clinical and scientific judgment when necessary, and moving ahead. I envy writers like Elisabeth Kolbert and others whose well-deserved stature enabled them to obtain support for their work, some of which has influenced mine. Kolbert was able to travel extensively and visit scientists as they worked. I traveled with my computer to the amazing library at the University at Buffalo where helpful librarians were usually able to provide me with papers from other repositories. It was rare to wait more than a day for a loan request to be fulfilled.

I am indebted to a great many individuals whose work I have relied on extensively during the course of writing this book. First and foremost are the authors of the peer-reviewed scientific publications that have provided what I hope is a solid, data- and evidence-based approach to my topic. Many of these authors have been extraordinarily helpful by giving permission to reproduce their work and sending me copies of papers not readily available at the University at Buffalo along with other relevant publications. The numerous scientists who contributed to the Intergovernmental Panel on Climate Change Fifth Assessment Reports have been a constant inspiration and unlocked the doors to important lines of inquiry that are not a usual part of a neurologist's training and experience. I am indebted to the often-nameless scientists who performed herculean work as they wrote, fact-checked, and reviewed reports published by governmental agencies such as the Environmental Protection Agency, the US Department of Agriculture, the US Energy Information Administration, and others. Although I have tried to rely on the peer-reviewed literature whenever possible, many outstanding organizations, including Physicians for Social Responsibility (PSR), the Sierra Club, and the Natural Resources Defense Council, have made important contributions. Earthjustice receives special thanks for its work and support.

Who can't help but be inspired by the following sources? In no particular order: James Balog, the genius behind the documentary *Chasing Ice*; Elizabeth Kolbert, author of *The Sixth Extinction*, a book that everyone should read; Al Gore, whose *An Inconvenient Truth* reminds us how to use the bully pulpit; and Bill McKibben and all the good people at 350. org. If James Hansen's voice had been heeded, this book would not have been necessary. There are scores of others that served as a source of ideas and inspiration.

During my rewarding and varied career as a physician–scientist, I learned to cherish the value of evidence-based decision-making. I try as hard as possible to be aware of sources of bias that affect thinking and behavior. I ask the same of others as we undertake the formidable task of planning for the future we want for our children, grandchildren, and the others who will follow us and who must live with the choices we make today. This generation may be the last one that has any hope of mitigating what some refer to as an impending climate change public health disaster or, from a decidedly more optimistic perspective, the opportunity to deal triumphantly with this public health opportunity. There are many win-win, no-regrets choices to make if we have sufficient wisdom to do so.

As I began final revisions of the book, Pope Francis visited the United States, where he addressed Congress and the United Nations. In his May 24, 2015, encyclical *Laudato si'*, he wrote: "Our sister [Mother Earth] now cries out to us because of the harm we have inflicted on her by our irresponsible use and abuse of the goods with which God has endowed her."[1] Five days later, a multinational group of health professionals wrote this about climate change in the preeminent journal *The Lancet*: "A healthy patient cannot continue with indefinitely rising levels of a toxin in the blood."[2] Religious leaders, climate scientists, and healthcare professionals all speak with a common voice.

I am proud to have been a member and supporter of Physicians for Social Responsibility for over three decades. I will donate all of the royalties from the sale of this book to PSR to help it in its mission to protect all of us from the greatest threats to survival.

Without the loving support of my wife, Anne, I could not have contemplated taking on the task of writing this book. She was always there, ready to provide the criticisms that improved my effort, find the mistakes that I failed to see, correct the spelling and usage errors that inevitably crept onto my pages (in spite of spell check), and bring me the odd cups of tea or coffee that helped sustain the effort needed to move ahead. It is almost inevitable that errors remain, in spite of Anne's able efforts and those of my editors at the MIT Press. Those mistakes are mine alone.

Alan H. Lockwood
Oberlin, Ohio

# 1

## Introduction

> Health is a state of complete physical, mental and social well-being and not merely the absence of disease or infirmity.
>
> —Preamble to the constitution of the World Health Organization, April 1948

When we saw the first pictures of the earth as viewed from space, it became evident that we are a small part of a system that is vast, complex, and interconnected. As scientists have learned more about the intricacy of these systems, we have discovered that life depends on their integrity. We barely understand some of the elements of even the simplest systems and are completely ignorant about altogether too many others. One of the fundamental goals of scientific, economic, geographical, and psychosocial research—as well as formal examinations of numerous other areas of study—is to better understand our world so that we may profit from this knowledge.

The law of unintended consequences tells us that intentional or unplanned incursions that involve these life-sustaining systems may produce an unforeseen result. To restate a point made by Donald Rumsfeld when he discussed the possibility of weapons of mass destruction in Iraq: "We also know there are known unknowns; that is to say we know there are some things we do not know. But there are also unknown unknowns— the ones we don't know we don't know. And if one looks throughout the history of our country and other free countries, it is the latter category that tend[s] to be the difficult ones."[1] Rumsfeld's admonition certainly applies to climate change, for which the consequences of complexity are dismissed by those who believe that it is a hoax and embraced by those who believe that it is real and happening right now.

To begin to understand the law of unintended consequences, extrapolate this simple example to the billions of interconnected systems that depend on the earth's climate. Three-toed sloths make a weekly, energy-consuming trip to the ground. Sloths have few enemies in the trees where they live. On the ground, predators are everywhere. What evolutionary advantage does this perilous trip confer on the sloths, and why do they make the trip? They make it to defecate. Research has shown that this act is a critical step in the sloth's life cycle.[2]

A unique species of moths that lives in the sloth's fur lays its eggs in the newly deposited dung: this is the only dung that will do. Newly hatched juvenile moths fly up to the forest canopy to find and live in the fur of the needed sloth. They then live out their entire lives in the sloth's fur, but the story does not end there. After the moths die, their bodies sustain algae that also live in the sloth's fur. These algae are the primary source of the sloth's nutrients; without the algae, the sloths would starve. An unintended disruption of this cycle could lead to the extinction of both the moths and the sloths.

The next example is perhaps more relevant to considerations of climate change. On September 4, 1882, Thomas Edison threw a switch that started the flow of electricity from the newly completed Pearl Street Station to John Pierpont Morgan's nearby office. Although the coal ash from the power plant quickly became a problem, there was no way that Edison could have known that the carbon dioxide produced by the coal that was burned in the power plant would start to change the climate. Now, more than a century and a quarter later, we know that burning coal and other fossil fuels injects billions of tons of carbon dioxide into the atmosphere each year. Scientific research has shown conclusively that this greenhouse gas is the most important factor driving climate change. Many who are dedicated to preserving and improving public health are convinced by the evidence that climate change poses an unprecedented threat to human health and the environment.

A recent report by the Lancet Commission on Health and Climate Change takes a more positive position, regarding the challenge we face as an unprecedented opportunity.[3] It is the totality of the environment that sustains the sloths and countless other species. Ultimately, the environment sustains all living creatures. Sloths are powerless to act. Are we also powerless? Or will we take advantage of the immense and diverse body of knowledge that we and those who have gone before us have worked so hard to acquire in order to act in our own self-interest?

## Ecosystems and Health

In 2000, then Secretary-General of the United Nations Kofi Annan called for a report that is now known as the Millennium Ecosystem Assessment. The portion of this report that focused on the relationships between human well-being and the environment was published in 2005.[4] In addition to evaluating the effects of changes in the ecosystem on health, the report also sought to place the need for sustainable uses of ecosystems on a firm scientific footing. Some of the report's conclusions about linkages between ecosystems and health are portrayed in figure 1.1. The report is based on several postulates:

- Ecosystems are the basis for the support of human and other forms of life.
- "Services" provided by ecosystems, such as providing food, water, and air, are a fundamental requirement for health, as defined by the World Health Organization at the beginning of this chapter.

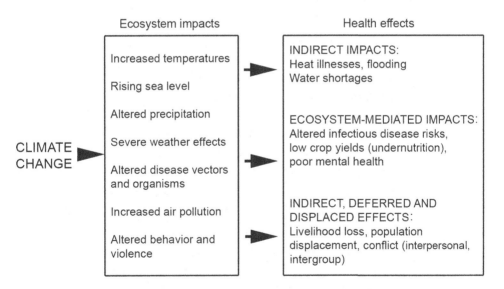

**Figure 1.1**

Critical relationships between climate change ecosystem effects and human health. Adapted from figure SDM 1 in C. Corvalán, S. Hales, A. J. McMichael, Millennium Ecosystem Assessment (Program), and World Health Organization, *Ecosystems and Human Well-Being, Health Synthesis: A Report of the Millennium Ecosystem Assessment* (Geneva: World Health Organization, 2005).

- The ecosystems that we depend on, as illustrated by the sloth/moth cycle, are complex because factors that modify them may be direct, indirect, or displaced in time, as illustrated by Edison's power plant.

A great deal of this book will be devoted to comprehensive discussions of how climate change affects the services provided by ecosystems and the impacts on health caused by the changes that have already occurred or that are likely to occur as the result of climate change.

### Aspirations for the Future

In 2000, the United Nations (UN) also sought to identify the most important goals for future development. In September of that year, with the support of numerous international organizations, all 183 members of the UN General Assembly adopted the Millennium Declaration. This unanimous action identified eight Millennium Development Goals (MDGs) that member nations hoped to achieve over the course of the next fifteen years. The goals were ambitious, many addressed health, and many are affected by climate change. They were

- to eradicate extreme poverty and hunger;
- to achieve universal primary education;
- to promote gender equality and empower women;
- to reduce child mortality rates;
- to improve maternal health;
- to combat HIV/AIDS, malaria, and other diseases;
- to ensure environmental sustainability; and
- to develop a global partnership for development.

The millennium is over, and progress toward achieving the MDGs was uneven but substantial, as summarized by the UN's 2015 report.[5] The hoped-for two-thirds reduction in childhood mortality did not occur, in spite of good progress. The member nations did not reach the goal for reduced maternal mortality. Too often, diseases such as malaria—one of many affected by climate—keep poor children from attending school, which impairs their ability to climb out of poverty, as described in more detail in chapter 5. Sadly, environmental sustainability is not yet within reach.

The environment and climate change are a major focus of the post-2015 UN efforts, as described by UN Secretary-General Ban Ki-moon, who wrote in his report to the General Assembly: "All voices have called for a people-centered and planet-sensitive agenda to ensure human dignity, equality, environmental stewardship, healthy economies, freedom from want and fear, and a renewed global partnership for sustainable development."[6] The Sustainable Development Goals successor to the MDGs includes seventeen objectives, including taking "urgent action to combat climate change and its impacts." The Secretary-General identified the following six essential elements that are required to meet sustainable development goals for the future:

1. Dignity: to end poverty and fight inequalities
2. Prosperity: to grow a strong, inclusive, and transformative economy
3. Justice: to promote safe and peaceful societies and strong institutions
4. Partnership: to catalyze global solidarity for sustainable development
5. Planet: to protect our ecosystems for all societies and our children
6. People: to ensure healthy lives, knowledge, and the inclusion of women and children

These goals have a familiar ring, but they have a more direct focus on climate change and the related goal of sustainable development.

Sustainable development depends on mitigating and adapting to climate change. This effort must begin with effective measures to curb the emission of greenhouse gases, particularly carbon dioxide. In spite of professed aspirations to limit carbon dioxide emissions, the nations of the world, including developed and undeveloped economies, continue to burn vast amounts of coal to generate badly needed electricity. This is a particular problem in China, the leading emitter of this gas; until recently, the unchecked combustion of coal was seen as being vital to China's expanding economy. In the United States, cries to "stop the war on coal" make it difficult for the Environmental Protection Agency (EPA) to exercise its authority under the Clean Air Act and curb carbon dioxide emissions by new and existing power plants, as described in the Clean Power Plan and related regulations.

But there is hope. In the run-up to the 2015 Conference of the Parties to the United Nations Framework Convention on Climate Change (COP21/CMP11), otherwise known as Paris 2015, the United States, China, Brazil, India, and other nations announced that they are poised to take the needed steps. It remains to be seen what will actually happen.

### Burden of Disease

To develop effective policies and allocate the resources needed to improve health, it is necessary to define a starting point. What are the factors that place health at risk? A project funded by the World Bank in 1990 attempted to answer this question. This report underwent a major update with the launch of a project known as the Global Burden of Disease Study 2010. The results of the efforts of a consortium of institutions were published in *The Lancet* in 2012. The first part of the report is a global assessment of over two hundred causes of death, broken down into different age groups, with a comparison to similar data from 1990.[7] Like its predecessor, this report evaluates what has already happened and draws few conclusions about the future.

A summary of selected elements of the Burden of Disease report is shown in table 1.1. Worldwide, ischemic heart disease is the leading cause of death. It is discouraging to see that this burden is about 35 percent higher than it was at the time of the 1990 report. Stroke follows in second place with a 25 percent increase. In spite of progress in combating HIV/ AIDS, there was a huge, almost 400 percent increase in deaths caused by the AIDS virus when compared to the 1990 data. Deaths due to tuberculosis and diarrhea both fell. Reported deaths due to malaria rose by just over 20 percent in the interval. Unfortunately, not all agencies or groups have used the same time frame for evaluating the malaria burden data. This has led to multiple answers and confusion. Malnutrition due to a lack of protein in diets, particularly among children, fell by just over 30 percent. Climate change is quite likely to have substantial effects on malaria and malnutrition, as discussed in subsequent chapters. Progress in the fight against undernutrition is a particular concern. The effects of higher temperatures and changes in rainfall already have made substantial impacts on the world's food supply.

**Table 1.1**

Selected causes of death

| Rank and cause | Mean rank (95% UI*) | Change from 1990 (95% UI*) |
|---|---|---|
| 1: Ischemic heart disease | 1.0 (1 to 1) | 35 (29 to 39) |
| 2: Stroke | 2.0 (2 to 2) | 26 (14 to 32) |
| 3: COPD | 3.4 (3 to 4) | -7 (-12 to 0) |
| 4: Lower respiratory infections | 3.6 (3 to 4) | -18 (-24 to -11) |
| 5: Lung cancer | 5.8 (5 to 10) | 48 (24 to 61) |
| 6: HIV/AIDS | 6.4 (5 to 8) | 396 (323 to 465) |
| 7: Diarrhea | 6.7 (5 to 9) | -42 (-49 to -35) |
| 10: Tuberculosis | 10.1 (8 o 13) | -18 (-35 to -3) |
| 11: Malaria | 10.3 (6 to 13) | 21 (-9 to 56) |
| 21: Protein energy malnutrition | 21.5 (19 to 25) | -32 (-42 to -21) |

*UI = uncertainty interval, the range of estimates that includes 95 percent of the expected answers computed from data used to define the cause of death.

The second part of the report on the Global Burden of Disease Study describes sixty-seven risk factors for the global burden of disease.[8] Risk factors, disability-adjusted life years (DALYs, or the sum of the years of life lost due to premature death and the years lost to disability), and worldwide deaths excerpted from this report are shown in table 1.2. Hypertension is the leading risk factor and has a clear causal link to ischemic heart disease and stroke. Although the report does not attempt to define the importance of air pollution in the pathogenesis of hypertension as well as other diseases, the link seems clear based on the conclusions of a committee of the American Heart Association and my 2012 book, *The Silent Epidemic: Coal and the Hidden Threat to Health*.[9]

Although enormous progress has been made in reducing hypertension-related deaths in the United States, the aforementioned report shows that such deaths remain a serious problem worldwide. Indoor and ambient small particle concentrations ($PM_{2.5}$, particles with an aerodynamic diameter less than 2.5 microns) are clearly identified as leading risk factors (see also chapter 7). Dietary deficiencies, many of which are or are potentially affected by climate, were also identified as substantial risk factors. Climate change is likely to be increasingly important as a driver of

Table 1.2

Risk factors for disease, worldwide

| Risk factor and rank | % change from 1990 | DALYs (1,000s) | Deaths (1,000s) |
|---|---|---|---|
| 1: Hypertension | 27% | 173,556 | 8,395,860 |
| 3: Indoor air pollution | -37% | 108,084 | 3,478.773 |
| 4: Insufficient fruit | 5% | 104,095 | 4,902,242 |
| 8: Childhood underweight* | -61% | 89,117 | 860,117 |
| 9: Ambient PM$_{2.5}$** pollution | 8.8% | 76,163 | 3,223,540 |
| 11+: Poor diet*** | Most increased by about 30% | 304,623 | 12,645,166 |
| 22: High processed meat | 22.1% | 20,939 | 840,857 |
| 23: Intimate partner violence | 23.8% | 16,794 | 186,365 |
| 26: Poor sanitation | Not available | 14,927 | 244,106 |
| 34: Unimproved water | Not available | 7,775 | 116,126 |

*The IPCC Fifth Assessment Report appears to subdivide this further under the category of undernutrition, manifesting as growth stunting (low height for age due to chronic undernutrition), wasting (low weight for height), and underweight (low weight for age).

**PM$_{2.5}$ indicates particulate pollution with aerodynamic diameter of 2.5 microns or less, those most damaging to human health.

***Poor diet includes multiple risk factors in order of risk level: high sodium, low nuts and seeds, iron deficiency, low whole grains, low vegetables, low omega-3 fatty acids, high processed meat, low fiber, vitamin A deficiency, zinc deficiency.

undernutrition, particularly among children, as will be described in subsequent chapters.

There was a great deal of good news in the Millennium Development Goal and Global Burden of Disease reports. Not nearly as many children less than five years of age are dying, deaths due to malaria and HIV are falling—particularly in the most recent years covered, and the toll of infectious diseases has fallen. In the most developed parts of the world, there has been substantial progress toward reducing the mortality associated with heart disease and cancer, which are still the two leading causes of death among Americans. As a result of all of these factors, the life expectancies for both men and women are increasing.

The Global Burden of Disease reports, like all others, are completely dependent on the methods employed by a large and diverse group of researchers. This can be illustrated by the results of a study of aging adults that concluded that there were 503,400 deaths due to Alzheimer's disease among Americans who were seventy-five years of age or older in 2010.[10] This toll is enormously higher than the estimate of approximately 84,000 deaths made by the Centers for Disease Control, and the resulting discrepancy would raise Alzheimer's disease from sixth to third place in the nation. The difference lies in a prospective study of defined cohorts in the high death study versus an analysis of death certificates in the low death study. The explanation centers on the fact that patients who have Alzheimer's disease often die from pneumonia. The underlying dementia may not appear in the official death records. Are there similar factors confounding the Global Burden of Disease and Millennium Development Goal studies? It is possible, but these are the best data available.

The Intergovernmental Panel on Climate Change (IPCC) Working Group II's contribution to its Fifth Assessment Report begins to introduce climate change into a consideration of the earth's burden of disease. The IPCC estimated that what it terms *climate-altering pollutants* were responsible for 7 percent of the global disease burden.[11] The full report by this working group paints a potentially bleak picture of the future world if unbridled emissions of greenhouse gases are combined with inadequate measures to improve energy efficiency and transition to sustainable sources of energy, along with a failure to make the adaptations needed to minimize the effects of rising oceans, more extreme weather events, and the higher temperatures that are likely.[12]

## The Trajectory toward the Future

Virtually all of us who believe that climate change is real, that human activity is largely the cause, and that prompt actions are badly needed to prevent future ramifications are dismayed by the failure of our institutions to take needed steps. We must move in a direction that is protective of human health and the environment, that helps to meet the Sustainable Development Goals, and that relieves pressure on the causes of morbidity and mortality enumerated by the Global Burden of Disease Study. The authors of a framework for approaching these issues list the following

items that must be addressed together to mitigate and adapt to our changing climate:[13]

- Political leadership needed to instigate and support the process
- Institutional organizations that can execute policy
- Extensive stakeholder involvement
- Climate change information—a primary objective of this book
- Use of decision analysis and decision-making tools
- Explicit consideration of barriers
- Funding
- Development and spread of needed technology
- Research

Examples of how elements of this schema apply to the Florida Keys and
the Netherlands are presented in chapter 10.

Without political leadership, the whole process fails. Little or no progress will be made. At the present time, sufficient political leadership is
absent. In the United States, President Obama's administration has taken
steps to mitigate climate change, primarily via rules (or *protections*, as I
prefer to call them) promulgated by the EPA. Many in the environmental
community regard these steps as too little, too late, and they have been
and will continue to be met by stiff and well-financed opposition. Parti-
sanship in the US Senate and particularly the House of Representatives is
at an all-time high. Not surprisingly, the nations with the most to lose and
the least influence have been the most adamant that something must be
done—and soon. President Tong of Kiribati has said, "Our people will
have to move." He expects his island nation to be inundated by a rising
sea level, as described in chapter 6.

Political will goes only so far. Detailed and specific plans must be created by institutional organizations that will accomplish defined objectives
designed to mitigate or adapt to climate change. Without adequate funds
and the institutional organizations needed to formulate policy into workable steps, nothing will happen. The steps most likely to be taken are
those that have been labeled as "no regrets" adaptations or mitigating
steps—those that can be justified under any possible circumstances,
including the absence of climate change.[14]

Psychologists have studied decision analysis in detail. To fail to use the best methods to make important decisions risks failure. Evidence-based decision analysis should be employed widely in making plans for the future. This is a critical focus of contemporary medical practice.

Finally, there must be strong and well-funded support for education as well as research and development programs. This is the only way to make the technological advances that will insure the ability of societies to cope with present and future challenges in the most efficient and appropriate manner.

# 2

## The Scientific Evidence for Climate Change

I trust that after what has been said the theory proposed in the foregoing pages will prove useful in explaining some points in geological climatology.

—Svante Arrhenius, 1896, from *On the Influence of Carbonic Acid in the Air upon the Temperature of the Ground*

Perched near the top of Mauna Loa, the world's largest volcano, Charles David Keeling in March 1958 made the first in a series of measurements of the concentration of carbon dioxide in the air. The average value for that month was 315.71 parts per million (ppm). This means that for every one million molecules in the atmosphere, 315.71 were composed of carbon dioxide. With the exception of a brief disruption in 1964, this record of timed measurements of the concentration of carbon dioxide in the atmosphere continues to the present day. This is one of the most important achievements and one of the most valuable sets of data from contemporary scientific research. On May 9, 2013, the average carbon dioxide concentration reached an unprecedented 400 ppm. This was the highest reading ever recorded at this site and the highest it has been in records that date back around twenty-four million years.[1] The data collected at the Mauna Loa laboratory have provided convincing scientific evidence that carbon dioxide–mediated global warming is upon us. This record is now known as the Keeling Curve.

The Mauna Loa site is an ideal location for measuring carbon dioxide and other atmospheric gases. These gases are referred to as being "well mixed," which means that there are few sources of the gases in the area and the prevailing winds that circle the globe mix gases quite evenly. In addition, there are vertical winds that mix gases to an extent between layers of the atmosphere that are close to the surface of the earth and those at high altitudes. In other words, the air samples obtained at Mauna Loa are a good representation of the average concentration of gases in the Northern Hemisphere.

Like many other notable scientists, Keeling experienced some difficult moments in his career. A few years after his initial seminal observations, the funding for his work from the National Science Foundation came to a halt. However, just a few years later, the Foundation cited his work in its 1963 warning about global warming. President George W. Bush awarded the National Medal of Science to Keeling in 2002, and Vice President Gore presented him with a special achievement award in 1997 on behalf of a grateful nation and as a recognition of the importance of his contribution.

Keeling's work, coupled with that of thousands of scientists who have studied the earth's climate, culminated in the 2013 report of the Intergovernmental Panel on Climate Change (IPCC), which states, "Warming of the climate system is unequivocal."[2] This is extraordinarily blunt language for scientists, who typically understate their conclusions and avoid controversy. The panel goes on to say that "since the 1950s, many of the observed changes are unprecedented over decades to millennia. The atmosphere and the ocean have warmed, the amounts of snow and ice have diminished, sea level has risen, and the concentrations of greenhouse gases have increased."[3] In spite of this detailed report prepared by literally hundreds of the world's leading climate scientists, many influential and well-financed individuals and organizations still exhort us to believe that climate change is a hoax. (Note that the list of contributors to the IPCC report is forty-seven pages long.)

### How Do We Know That the Climate Has Changed?

Medical education begins with the study of first normal, then pathological biochemical and physiological processes and anatomical structures. A student must learn what is normal before tackling and dealing with the abnormal. The same is true for the study of climate. The term *climate change* indicates that there has been a deviation from a prior state. In the field of climate science, as in medicine, it is important to understand what has changed and why before effective actions can be taken. In a second parallel with medicine, actions designed to prevent climate change are virtually certain to be preferable to those needed to deal with a new and arguably pathological state.

It is important to distinguish between *weather* and *climate*. Weather is a precise description of specific elements of the atmosphere, such as temperature, humidity, and wind speed and direction at a given time and place. Climate is a description of the weather over an extended period of time, often years. An understanding of climate and climate change depends on an understanding of what determines climate and what affects the forces that determine climate. The scientific literature commonly refers to these forces as the drivers of climate. The task of understanding climate begins with an analysis and understanding of the earth's energy budget.

Your personal household budget reflects income from salary, interest, and other sources, along with expenditures for food, housing, transportation, and so on. The earth's energy budget is analogous, but like the budget of a nation, it is complex. However, this budget can be simplified to an extent. Energy from the sun is deposited on the earth in the form of solar radiation or light at different wavelengths. Different wavelengths of light have different properties. For example, infrared wavelengths transfer heat. Some of the light energy that reaches the earth is reflected back into space by clouds, the atmosphere, snow and ice on the earth's surface, and so on. Light that reaches the earth's land areas and oceans may be absorbed and converted into heat energy that warms the planet. Some of the heat from the warmed earth radiates back toward space in the form of long wavelength infrared energy. Some of that heat escapes into space, but some is absorbed by gases, specifically the greenhouse gases that are in the atmosphere. This absorption of energy heats the layer of the atmosphere where this process takes place, and these greenhouse gases then reemit heat energy in every direction. Some of this heat energy returns to the earth, warming it. This is the greenhouse effect.

The amount of heat energy trapped by greenhouse gases is determined by the concentration of the gas and its physical properties. Not all greenhouse gases are equal. Different gases have different properties that make them more or less efficient at trapping heat. The amount of trapped heat increases as the gas concentration increases. In terms of the earth's energy budget, an increase in greenhouse gas concentration reduces the amount of heat energy that escapes from the earth. From a budgetary perspective, the input of heat remains the same, but the output of heat decreases as more greenhouse gases are added to the atmosphere. In accord with the laws of physics, this extra heat energy is distributed throughout the air, land, and water: they all become warmer.

The scientists who study the climate and produced the IPCC reports use the term *radiative forcing* (RF) to describe the change in downward minus upward energy transfer to and from the earth at the top of the atmosphere that is due to a driver of climate change. Positive RF leads to warming of the earth and negative RF to cooling of the earth. Each component of the atmosphere makes its unique contribution to the total value for RF. The contribution to the total RF made by each greenhouse gas can be determined with great accuracy. By using information about the atmosphere of the past, present-day values for RF can be compared to past RF values. Changes in the RF of the earth can be due to human activity or due to natural phenomena. For example, greenhouse gases—such as carbon dioxide, methane, and nitrous oxide—all act to increase the total RF. Volcanic eruptions that inject large amounts of dust and sulfur dioxide into the upper atmosphere decrease the total RF. Transient periods of cooling have been observed after particularly large eruptions. The current value for RF is about 2.3 watts per meter squared ($W/m^2$) higher for the whole earth compared to the 1750 pre–Industrial Revolution value. The overwhelming majority of climate scientists agree that this increase is due to human activity and is driving climate change. Disagreement among scientists centers on uncertainties in the data due to measurement error (error is an element in every measurement) and the concentration of relevant atmospheric components.

### The Temperature Record

Changes in the earth's temperature are at the heart of the argument that the earth's climate is warming.

The ability to measure the temperature of the earth, or any other object, is something we take for granted now, but this has not always been the case. The concept that the volume of air or other substances changed when heated or cooled was known in ancient times. Primitive instruments called *thermoscopes* were built by a number of early scientists, including Galileo. They consisted of a tube partially filled with air and water. The air–water boundary moved as the instruments were heated or cooled and as the air pressure changed. What we would recognize as a modern thermometer was born when developers filled and sealed a liquid in a bulb-tube device and applied a reference scale. This made it possible to make and compare measurements that were not also partly dependent on air pressure. The earliest measurements of the temperature of the atmosphere began in the middle of the seventeenth century.

The earliest continuous air temperature measurements were made in the Midlands of England, beginning around 1659. These early data show that the average temperature in January was 3°C, with a yearly average of 8.8°C. These data are shown in panel A of figure 2.1. Although these early measurements lack the precision and accuracy that are characteristic of the thousands of measurements made all over the world with contemporary instruments, the values recorded over the next 314 years show little deviation from an average yearly temperature of around 9°C, which is about 49°F. This is a marked contrast to the data shown in panel B of figure 2.1, showing a substantial rise in the global surface temperature that began around the start of the twentieth century. Thus, there was little if any change in the temperatures recorded in the Midlands of England between 1659 and 1973. This stands in contrast to the increase in the global surface temperature of about 0.8°C that has taken place between 1900 and the early part of this century.

Although the instrumental age of temperature recordings began in the mid-seventeenth century, the earth's temperature record reaches much farther back into the past. One remarkable study yielded a temperature record that extends back almost six hundred million years.[4] These data point to temperatures that were around 7°C warmer than they are now. Other less dramatic but more direct records extend back just over 420,000 years, as shown in figure 2.2.

In order to compute temperatures at these extreme dates, scientists have developed so-called proxies for the measurement of the actual temperature with a thermometer. Proxies depend on well-preserved physical aspects of samples that can be analyzed under laboratory conditions and related, in a verifiable manner, to the temperature of the sample when it formed. Examples of proxy measurements include very old ice samples, growth rings in trees, pollen grains, and samples of coral or other marine animals. Some of the most useful proxy data have

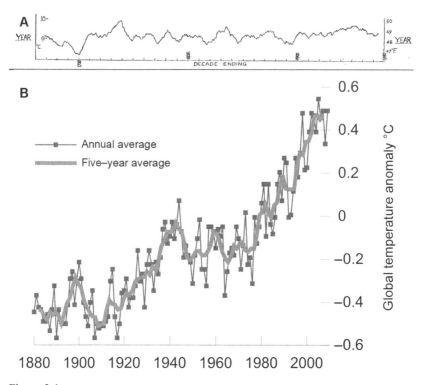

Figure 2.1

A: Monthly average temperatures recorded in the Midlands region of England
from 1659 to 1973 showing relative stability during this period. G. Manley,
"Central England Temperatures: Monthly Means 1659 to 1973," *Quarterly Journal of the Royal Meteorological Society* 100 (1974): 389–405. B: Global near-
surface air temperatures from 1880 to 2010, compiled by the NASA Goddard
Institute for Space Studies. The zero point is the average temperature from 1961–
1990, as per the practice of the IPCC. The figure shows a general warming trend
during this time. Similar data have been published by NOAA, the Hadley Unit of
the UK Meteorological Office, and Berkeley Earth. Figure prepared by Robert A.
Rhode for the Global Warming Art project, reproduced per terms of GNU free
documentation license v 1.2.

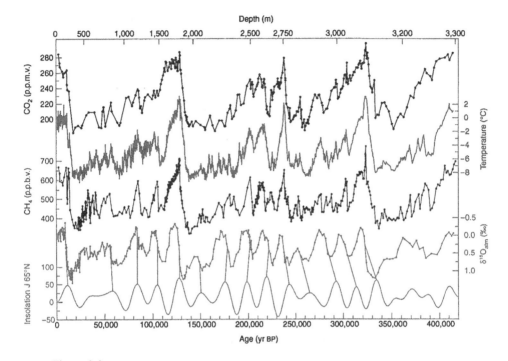

Figure 2.2

Climate data from Vostok ice cores covering 420,000 years. From top to bottom: atmospheric $CO_2$ concentration in parts per million by volume, which reached 400 ppmv in 2013; temperature change determined from hydrogen isotopes; $CH_4$ concentration in parts per billion by volume; changes in global ice volume determined from oxygen isotopes; and solar energy deposits at 65 degrees N latitude, in joules. Figure rendered from color image in Wikimedia Commons file Vostok 420ky 4 curves insolation, in public domain.

come from core samples obtained by drilling into Antarctic ice sheets and sediments at the bottoms of oceans. These core samples contain air bubbles that yield important information about the composition of the ancient atmosphere.

To better understand proxy measurements that extend far into the past, research must account for fluctuations in the amount of solar energy that reaches the surface of the earth (solar insolation). These changes are due to the well-known characteristics of the earth's orbit around the sun and include the following:

- Changes in the shape of the earth's orbit (technically, the orbital eccentricity, with two different periods of about 100,000 and 400,000 years)
- Variations in degree to which the earth tilts on its axis (obliquity, with a period of about 42,000 years)
- The tendency for the earth to wobble on its axis, like a top (axial precession, with a period of 23,000 years)

These changes in the earth's orbit are called Milankovitch cycles. Some of the impacts of these cycles on climate data are shown in figure 2.2.

Changes in the amount of solar energy reaching the earth due to predictable variations in the orbit of the earth occur extremely slowly—over the course of thousands of years. The rate of temperature changes due to orbital fluctuations is too gradual to play any role in the climate changes taking place now.

The ingenuity of modern scientists was taxed as they developed and validated proxies for actual measurements of temperature. Widely used techniques depend on measuring differences in the mass ratio or weight of molecules that contain different isotopic forms of elements, such as oxygen and hydrogen. Most oxygen exists in one of two stable (nonradioactive) forms: oxygen-16 ($^{16}O$ ) and oxygen-18 ($^{18}O$). The nucleus of each form contains eight protons; this makes it oxygen. Oxygen-18 contains ten neutrons, whereas $^{16}O$ contains only eight, making $^{18}O$ water slightly heavier than its $^{16}O$ counterpart. This difference in weight between water molecules that contain $^{18}O$ versus $^{16}O$ causes slight but definable differences in the way these water molecules behave at different temperatures. The lighter $^{16}O$ water molecules evaporate more readily than their heavier counterparts. Similarly, the heavier $^{18}O$ water molecules condense out of the atmosphere before the lighter molecules.

By measuring the ratio of $^{18}O$ to $^{16}O$ in an ice-core sample and comparing this ratio to an appropriate reference sample, it is possible to make accurate deductions about the temperature, the amount of rainfall, or the amount of ice covering the earth when the sample was fixed by freezing (see box 2.1). A 420,000-year record of surface temperatures at the Vostok site in Antarctica and global ice coverage made using oxygen isotopic techniques is shown in figure 2.2. During the period when this ice formed, temperatures were typically lower than at present. However,

**Box 2.1**

Since $^{16}$O is lighter than $^{18}$O, water containing the lighter isotope evaporates more readily than water containing the heavier isotope. Water containing heavy oxygen condenses out of the atmosphere more readily than water containing light oxygen. Water vapor from the warm equator, high in $^{16}$O, tends to move toward the cooler polar regions of the earth. This increases the amount of $^{18}$O in equatorial waters.

On the way toward the poles, the air cools to form precipitation that preferentially depletes the amount of $^{18}$O in the air. When the remaining water vapor reaches polar regions, it contains much less heavy oxygen and more light oxygen. This increases the amount of light oxygen in polar precipitation. High concentrations of the heavier oxygen in oceans recovered from oceanic core samples, trapped as carbonate in the shells of sea creatures such as foraminifera, are characteristic of periods when large amounts of ice covered the earth. Ice cores from polar regions with low amounts of $^{18}$O indicate low polar temperatures. When the climate warms, polar ice (rich in $^{16}$O) melts, reducing the salt content of ocean water and the $^{18}$O to $^{16}$O ratio. Thus, by measuring the isotopic forms of oxygen in various samples and comparing this ratio to standards, proxy measurements of temperature and ice coverage of the earth can be deduced.

This isotopic ratio technique has other applications. Nonradioactive carbon isotopes ($^{12}$C and $^{13}$C) are used to determine whether atmospheric carbon dioxide is derived from burning fossil fuels.

occasional peaks of about 2°C above current temperatures occurred that correspond to the warm periods present in the intervals between ice ages. These peaks are correlated with peaks in the concentration of carbon dioxide ($CO_2$) and methane ($CH_4$) in the atmosphere. These gas concentration data are the result of analyses of the small air bubbles that were trapped in the ice when they formed. A similar core drilled by the European Project for Ice Coring in Antarctica (EPICA) extends back about 801,000 years.

The data obtained from analyses of the air bubbles trapped in Antarctic ice laid down over thousands of years show that substantial temperature changes have occurred in the past. However, these changes occurred much more slowly than the changes measured since the beginning of the Industrial Revolution that continue through to the present.

Critically, as discussed ahead, the concentrations of greenhouse gases in the atmosphere were much lower when the ice formed than they are at the present.

Thus, it is possible to conclude that the earth is warming—and at a rate that is greater than any rate measured for the past 800,000 years. As a corollary to the temperature data, the present concentration of greenhouse gases that are driving climate change exceeds any concentrations found during those hundreds of millennia, and they are rising at a rate that is faster than in times past.

### Long-Lived Greenhouse Gases

"It is *extremely likely* that human influence has been the dominant cause of the observed warming since the mid-twentieth century. This is evident from the increasing greenhouse gas concentrations in the atmosphere."[5] In this sobering conclusion published in the most recent IPCC report, "extremely likely" is shorthand for a 95 to 100 percent probability that the statement is true. These increases have occurred in spite of several high-profile international conferences, such as those that took place in Kyoto and Rio de Janeiro, where participating nations failed to produce a binding agreement that would reduce greenhouse gas emissions. A 2013 report by the United Nations Environment Programme estimated that global greenhouse gas emissions in 2005 were the equivalent of forty-five billion metric tons (45 gigatons [Gt], where 2,200 lb = 1 tonne) and rose to 49 Gt by 2010.[6] The UN's business-as-usual scenario predicts a further rise to 59 Gt by 2020. Most of the current relentless increase is driven by burning fossil fuels in developing nations, China, and the United States.

Scientists who have studied the greenhouse effect and its link to climate change have shown clearly that there is a net gain in the earth's heat energy (positive RF) since the dawn of the industrial age—arbitrarily established as 1750. Large increases in the atmospheric concentration of greenhouse gases, particularly those that remain in the atmosphere for long periods, have acted as the primary drivers for the increase in RF. The principle greenhouse gases are carbon dioxide, methane, nitrous oxide, stratospheric ozone-depleting halocarbons and their substitutes (chlorofluorocarbons, hydrochlorofluorocarbons, hydrofluorocarbons, perfluorinated carbon compounds, and other chlorinated and brominated chemicals), and sulfur hexafluoride. Carbon dioxide, methane, and nitrous oxide occur naturally but also are emitted as the result of human activity. The remainder—that is, all of those containing chlorine or fluorine—are all products of industrial activity. Serial measurements using sensitive instruments have provided the data to support this conclusion. Some data are the result of atmospheric sampling, such as that which Charles Keeling started in the late 1950s, whereas other data have come from an analysis of air bubbles trapped in ice cores.

## Carbon Dioxide

Carbon dioxide makes the largest contribution to global warming of all of the long-lived greenhouse gases. This is due to its high concentration, its heat trapping ability (RF; see table 2.1), and the very long period for which newly formed $CO_2$ remains in the atmosphere. A 2005 study reported that the average lifetime for atmospheric $CO_2$ is between thirty thousand and thirty-five thousand years and that between 17 and 33 percent of the $CO_2$ in the air today will still be in the atmosphere one thousand years from now.[7] This long lifetime has led climate scientists to use $CO_2$ as the basis for comparing the global warming potential of all of the long-lived greenhouse gases (see table 2.1).

Table 2.1

Properties of long-lived atmospheric global warming gases (IPCC Fifth Assessment Report, Working Group I, Table 8.2)

| Gas | Concentration, 2011 | Lifetime (years) | Radiative Forcing, 2011 $(W/m^2)$ | 20-year GWP | 100-year GWP |
|-----|---------------------|------------------|-----------------------------------|-------------|--------------|
| $CO_2$ | 391 ppm | 30,000–35,000 | 1.82±0.19 | 1* | 1* |
| $CH_4$ | 1,803 ppb | 12.4 | 0.48±0.05 | 86 | 34 |
| $N_2O$ | 324 ppb | 121 | 0.17±0.03 | 268 | 298 |

*The global warming potential for $CO_2$ is set to one by definition.
Abbreviations: W = watts, m = meter, ppm = parts per million, ppb = parts per billion, GWP = global warming potential relative to $CO_2$. (*Note:* These values include the effects of climate-carbon cycle feedback, the influence of climate change on the earth's carbon cycle. Exclusion of this feedback yields slightly smaller values.)

In more technical terms, total RF in the time interval between 1750 and 2011 attributable to human activity is around 2.29 W/m$^2$ (90 percent of the estimates range between 1.13 and 3.33 W/m$^2$). Approximately 73 percent of that total, or 1.68 W/m$^2$, is attributed to $CO_2$ according to the IPCC Fifth Assessment Report (90 percent of these estimates range between 1.33 and 2.33 W/m$^2$).

Carbon is present in all organic molecules and is essential for life. Because it is found everywhere and in all living things, it is not surprising that the earth's carbon budget is extremely complex. Figure 2.3 is a simplified overview of the carbon cycle that depicts the equilibrium between the atmosphere, the oceans, and terrestrial systems, and how human activity has disrupted the carbon cycle.

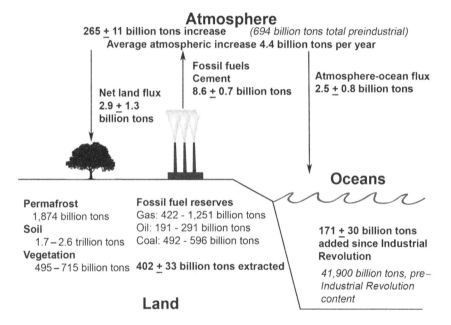

**Figure 2.3**

The earth's carbon cycle. This is a highly simplified version of the IPCC Working Group I, Fifth Assessment Report, Figure 6.1. It shows carbon movements between the atmosphere, land, and oceans, omitting many details. All values are in US tons of carbon (*not* tons of $CO_2$). Fluxes, indicated by arrows, are tons per year at the present time. Bold face type indicates changes after the year of 1750. Regular type indicates fossil fuel and permafrost reserves.

The ability to make serial measurements of the atmospheric concentration of $CO_2$ has led to valuable insights into the mechanisms causing climate change. The record of these measurements, now referred to as the Keeling Curve, is shown in figure 2.4. The curve has two essential features: a seasonal fluctuation, superimposed on a relentless increase. Work based on Keeling's measurements has shown that the reductions in the concentration that occur during the spring and summer are due to carbon dioxide that is trapped in the new growth of trees and other plants in the Northern Hemisphere. In the fall and winter, when this growth has ceased, the concentration rises.

Data from ice cores, shown in figure 2.2, show the concentration of $CO_2$ in the atmosphere for the past 420,000 years. There have been five $CO_2$ concentration peaks in that interval, which are referred to as the

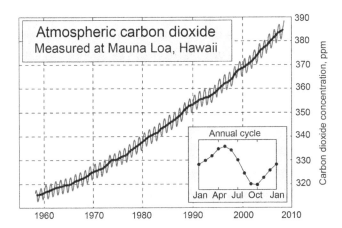

**Figure 2.4**

The Keeling Curve. This is the record of the concentration of carbon dioxide in the atmosphere at the Mauna Loa Observatory from the late 1950s to the present. It shows a steady increase in the concentration of this important greenhouse gas and the annual fluctuations. The decreases in the concentration during the Northern Hemisphere spring are due to plant growth and carbon dioxide sequestration. In the fall, this process ends, and carbon dioxide levels rise during the winter months. Source: Wikimedia Commons, not copyrighted.

periods of interglacial warming. As seen in the figure, the highest concentration of $CO_2$ from the ice core data is approximately 300 ppm. A concentration of 400 ppm was recorded on May 9, 2013. Since 1980, the concentration of $CO_2$ has risen at an average rate of 1.7 ppm per year. It is certain that the cyclic rise described by Keeling will continue into the foreseeable future.

How can we be so certain that this excess of $CO_2$ is due to human activity? Climate change deniers point to the natural $CO_2$ variability shown in figure 2.2 and contend that the recent changes are nothing new—but they are wrong. The current $CO_2$ level is something quite new. The present concentration is higher than at any time in the last twenty-four million years.[8]

The link between burning fossil fuels and the rise in atmospheric carbon dioxide is firm. Records and estimates of fossil fuel combustion correlate well with the rate and magnitude of the rise in the atmospheric $CO_2$.

In chemistry, we learned that $CO_2$ is formed by the creation of the chemical bond between an atom of carbon and a molecule of oxygen:

$$C + O_2 \rightarrow CO_2 + heat$$

From this equation, we can see that a molecule of oxygen is consumed when a molecule of $CO_2$ is created. Because the concentration of oxygen in the atmosphere is very high, it is extremely difficult to measure the very small changes in its concentration predicted by the equation. Nevertheless, this is what Keeling's son did. His work has shown that there is a cyclic change in the concentration of atmospheric oxygen that matches the prediction made by his father's work.[9] These data are a second critical part of the proof that the increase in the concentration of $CO_2$ in the atmosphere is due to burning fossil fuels and not the result of natural processes.

Advances in analytical chemistry and a better understanding of photosynthesis have made it possible to produce one of the most sophisticated pieces of evidence linking burning fossil fuels with the rise in atmospheric $CO_2$.[10] Like oxygen, carbon exists in several isotopic forms. Carbon-12, containing six protons and six neutrons, accounts for about 99 percent of the element. The remaining 1 percent contains an extra neutron, making it $^{13}C$. Both are nonradioactive. (The very tiny amounts of radioactive $^{14}C$

remove it from consideration in a geological time perspective because of its half-life of about 5,700 years.) Carbon dioxide containing the lighter $^{12}$C enters the leaves of plants more easily than its heavier counterpart. That, along with the fact that certain photosynthetic pathways prefer the lighter isotopic form, leads to a dilution of $^{13}$C in plants compared to the $^{13}$C in the atmosphere. When burned, the resulting $CO_2$ has less of the heavier isotopic form and more of the lighter isotope. Because fossil fuels are derived from prehistoric plants, the $CO_2$ from fossil fuel combustion is also "plant-like" in terms of its $^{13}$C to $^{12}$C ratio when compared to a standard. This low $^{13}$C to $^{12}$C ratio allowed scientists to identify fossil fuels as the source of the increased atmospheric $CO_2$. Serial measurements of this ratio between 1980 and 2002 show a progressive decline, the result expected as a consequence of burning plant-based fossil fuels.[11] The fall in the isotope ratio also parallels the increase in fossil fuel consumption during this same period.

### Methane

Of the long-lived global warming gases, $CH_4$ makes the second most important contribution to climate change. As shown in table 2.1, the global warming potential of methane is eighty-six and twenty-four times greater than that of $CO_2$ when measured twenty and one hundred years after emission, respectively. The concentration of methane in the atmosphere has risen fairly steadily since the beginning of the Industrial Revolution, as shown in figure 2.5. In the early years of this century, there was a period of apparent stabilization followed more recently by a return to an upward trend.[12] Although the reasons for the stabilization are not completely clear, the resumption of the increase is thought to be due to emissions from the fossil fuel sector (oil, coal mines, coal-bed methane, and natural gas). This increase and other considerations have apparently triggered the plans to regulate methane emissions that the EPA announced in the early part of 2015.

The global methane budget is complex due to multiple incompletely understood natural and anthropogenic sources, shown in part in table 2.2. Total methane emissions from anthropogenic sources in the interval between 2000 and 2009 are thought to be around 365 million tons. The fossil fuel industry is responsible for about a third of all anthropogenic methane emissions (table 2.2). Natural sources emit about 382 million

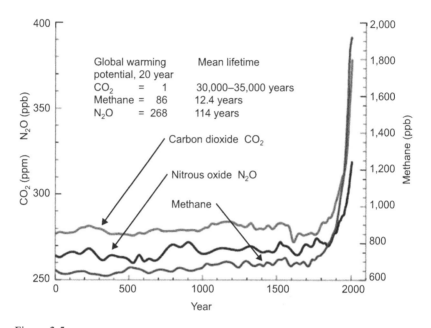

**Figure 2.5**

Atmospheric concentrations of important long-lived greenhouse gases over the last 2000 years. Increases since about 1750 are attributed to human activities in the industrial era. Concentration units are parts per million (ppm) or parts per billion (ppb), indicating the number of molecules of the greenhouse gas per million or billion air molecules, respectively, in an atmospheric sample. Modified from an original IPCC figure (FAQ 2.1, Figure 1, IPCC Fourth Assessment Report) and reproduced in accord with their copyright requirements.

tons. Total methane sinks are estimated to be around 666 million tons per year. Most atmospheric methane is oxidized in the lower atmosphere. Unfortunately, ground-level ozone is produced by these reactions, producing secondary effects on climate and health. There are substantial uncertainties in this budget due to emerging data that seek to quantify fugitive methane leaks at every stage of natural gas production, distribution, and utilization. There is also uncertainty about the fate of huge amounts of methane trapped in water crystals in the permafrost. These ice-like crystals known as *methane clathrates* can be burned. Some call clathrates "burning ice," and they form a portion of the carbon trapped in the permafrost shown in figure 2.3.

Table 2.2

Global methane budget

| Decade/Source | 1980–1989 | 2000–2009 |
|---|---|---|
| **Natural sources** | 391 (269–514) | 382 (262–534) |
| Wetlands | 225 (202–293) | 239 (195–313) |
| Other | 143 (67–220) | 143 (67–220) |
| Termites | 12 (2–24) | 12 (2–24) |
| Geological, oceans | 60 (36–83) | 60 (36–83) |
| Hydrates | 6.6 (2.2–9.9) | 6.6 (2.2–9.9) |
| Permafrost | 1.1 (0–1.1) | 1.1 (0–1.1) |
| **Anthropogenic sources** | 340 (322–356) | 365 (335–406) |
| Agriculture and waste | 204 (190–217) | 220 (206–247) |
| Ruminants | 94 (89–99) | 98 (96–104) |
| Landfills | 61 (55–66) | 83 (74–99) |
| Biomass burning | 37 (34–41) | 39 (35–43) |
| Fossil fuels | 98 (98–98) | 106 (94–116) |
| **Loss** | 594 (453–740) | 666 (532–814) |
| **Total Arctic carbon** | | 1,843 billion tons |
| **Carbon trapped in permafrost** | | 1,615 billion tons |

*Note:* Unless otherwise noted, units are in millions of US tons (2000 lb/ton) and are taken from Table 6.8 of *Climate Change 2013: The Physical Science Basis; Contribution of Working Group I to the Fifth Assessment Report of the Intergovernmental Panel on Climate Change*, ed. T. F. Stocker, D. Qin, G-K. Plattner, et al. [New York: Cambridge University Press, 2014]. Numbers in parentheses represent ranges from sources included in the table. Data for Arctic carbon are from C. Tarnocai, J. G. Canadel, E. A. G. Schurr, P. Khury, G. Mazhitova, and S. Zimov, "Soil Organic Carbon Pools in the Northern Circumpolar Permafrost Region," *Global Biogeochemical Cycles* 23, no. 2 (2009), doi:10.1029/2008GB003327.

Issues surrounding methane (the principal component of natural gas) and its role in global warming have taken on new dimensions as the result of arguments suggesting that natural gas can serve as a "bridge" from coal to more sustainable sources that make smaller contributions to global warming. The crux of the argument centers on the fact that burning methane produces about 57 percent of the amount of $CO_2$ per unit of heat produced compared to coal. Burning coal has numerous and serious adverse health effects, strengthening the bridge-fuel argument.[13] The recent boom in natural gas extraction, particularly with the widespread use of hydraulic fracturing, or fracking, has led to a substantial reduction in the price of natural gas. Coupled with increasingly stringent emission requirements for coal-fired power plants and pending rules that will regulate carbon dioxide emissions by new and existing power plants, this price reduction has triggered a shift from coal to natural gas in many electricity-generating units. Alas, the solution is not that simple.

The amount of methane that escapes into the atmosphere from the natural gas industry is contested hotly. Several recent studies illustrate the debate. Scientists sponsored by the National Oceanographic and Atmospheric Administration (NOAA) mapped methane plumes over gas wells in Uintah County, Utah, in 2012.[14] They reported that between 6.2 and 11.7 percent of all of the natural gas produced in the field studied escaped into the atmosphere. Using similar technology contained in a motor vehicle, others mapped all of the streets in Boston, Massachusetts.[15] They identified 3,356 sites where methane appeared to be leaking into the atmosphere. In some places, methane concentrations were approximately fifteen times greater than the concentration in well-mixed greenhouse gas samples, such as those collected at multiple sites in the Northern Hemisphere. A similar study of Washington, DC, identified almost six thousand leaks, with some methane concentrations reaching explosive levels.[16] These investigators measured the $^{13}C$ to $^{12}C$ isotopic ratio in their methane samples and concluded that the methane came from the natural gas distribution system as opposed to natural sources. Similar studies have been performed in California, with an emphasis on the Los Angeles area, and Indianapolis, Indiana.[17] On the basis of these studies, it seems likely that the estimate of the fossil fuel industry's contribution to anthropogenic sources of methane, as shown in table 2.2, is too low.

This possibility is supported by another recent study of methane releases throughout the United States. This study concluded that the estimates made by the US EPA and the Emissions Database for Global Atmospheric Research (EDGAR) were low by factors of about 1.5 and 1.7, respectively.[18] When various sources were estimated, these investigators concluded that the methane emissions from cattle (ruminants and manure) may be twice as high as suggested in prior studies. The prior estimates were particularly low for the south-central portion of the United States. In that region, the authors of this new methane study found that emissions were about 2.7 times greater than reported by others. Fossil fuel extraction and processing in refineries were identified as contributing to almost half of this excess, leading to the conclusion that fossil fuel emissions were almost five times greater than reported in the comprehensive EDGAR database. Emissions from three states—Texas, Oklahoma, and Kansas—account for 24 percent of all US methane emissions.

Although methane leaks from wells are suspected as a source for a substantial amount of methane, there is little consensus on this point. In a widely heralded study sponsored in part by the natural gas industry and the Environmental Defense Fund, investigators from several top-tier universities reported that emissions during natural gas production were approximately 0.42 percent of the total yield. This finding was challenged by Anthony Ingraffea, a Cornell University professor widely regarded for his expertise in this field, and his colleagues. He pointed out several potential sources of bias in the study, including the small number of wells using hydraulic fracturing techniques that were industry-selected for study.

In numerous public appearances, Ingraffea has reported data from publicly available sources detailing methane leaks from wells drilled in Pennsylvania. These data were published in the highly respected Proceedings of the National Academy of Sciences in the summer of 2014.[19] Ingraffea's team found that methane leaks from wells that took advantage of horizontal drilling with hydraulic fracturing techniques are several times more common than methane leaks from wells drilled in a more conventional vertical manner. They also reported that leaks from wells drilled after 2009 are more common than leaks from older wells. Because the location of many abandoned wells is not known precisely, measuring leaks from such wells is difficult. Thus, the true magnitude of the leaking well problem may be difficult to determine. In the meantime, approaches such as more rigorous controls of drilling procedures and better attention to constructing well casings that are more durable and more likely to remain intact and not leak are warranted.

The magnitude of the methane leak problem has enormous implications for combating global warming. Even though its lifetime in the atmosphere is short relative to $CO_2$, the global warming potential for methane is high, as shown in table 2.1. Theoretical projections of methane releases that have since been essentially confirmed indicate that the impact of methane on global warming may be greater than that of coal when fugitive methane releases are added to the $CO_2$ produced by burning natural gas.[20] This forms the crux of Ingraffea's argument that natural gas is a "gangplank to a warm future," as he wrote in an op-ed in the *New York Times*.[21]

The IPCC's methane budget estimates include emissions from permafrost and methane hydrates or clathrates (see table 2.3 and figure 2.3). Their data summary suggests that the total amount of carbon trapped in permafrost is between 1.5 and six times the amount in natural gas reserves. In the Fifth Assessment Report, these scientists also state that it is *virtually certain* (99–100 percent probability) that climate change will cause a retreat of the permafrost, which will release methane. In a 2011 report in *Nature*, scientists from the Permafrost Network wrote that the magnitude of carbon releases from thawing of the permafrost is "highly uncertain" but "[their] collective estimate is that carbon will be released more quickly than models suggest, and at levels that are cause for serious concern."[22]

### Halocarbons and Related Compounds

*Halocarbons* are carbon-containing molecules with one or more molecules of fluorine or chlorine. These chemicals do not occur in nature; they all are the result of human activity and industrial processes. The original compounds were developed in the late nineteenth century, but they were not synthesized in large quantities until the late 1920s and 1930s, when they replaced other hazardous chemicals used as refrigerants. Widespread use for this purpose, as propellants in spray cans, and in the electronics industry followed thereafter.

In the late 1970s and early 1980s, atmospheric scientists found that these gases were responsible for the destruction of the stratospheric ozone layer. Stratospheric ozone blocks ultraviolet (UV) radiation from the sun. Without this protective layer, many plants would not be able to grow. Other adverse effects caused by UV radiation, including dramatic increases in skin cancers, were predicted. As a result, the Montreal Protocol on Substances that Deplete the Ozone Layer was adopted, and recovery of the ozone layer is under way.

Then President Ronald Reagan, not known for his support of environmental regulations, was one of the champions of this treaty, which some have referred to as "the Little Treaty That Could." Similar chemicals that do not affect the ozone layer were developed to replace banned

Table 2.3
Predicted global mean sea level increases relative to 1986–2005 for four relative concentration pathway (RCP) scenarios

| Source/Scenario | RCP2.6 | RCP4.5 | RCP6.0 | RCP8.5 |
| --- | --- | --- | --- | --- |
| Oceanic thermal expansion relative to 1986–2005 (m) | 0.14 (0.10–0.18) | 0.19 (0.14–0.23) | 0.19 (0.15–0.24) | 0.27 (0.21–0.33) |
| Melting glaciers relative to 1986–2005 (m) | 0.10 (0.04–0.16) | 0.12 (0.06–0.19) | 0.12 (0.06–0.19) | 0.16 (0.09–0.23) |
| Rise: 2081–2100 relative to 1986–2005 (m) | 0.40 (0.26–0.54) | 0.47 (0.32–0.62) | 0.47 (0.33–0.62) | 0.62 (0.45–0.81) |
| Rate of rise 2081–2100 (mm/year) | 4.3 (2.0–6.6) | 6.0 (3.5–8.5) | 7.3 (4.6–10.0) | 11.1 (7.4–15.5) |
| Meters rise 2046–2065 relative to 1986–2005 | 0.24 (0.17–0.31) | 0.26 (0.19–0.33) | 0.25 (0.18–0.32) | 0.29 (0.22–0.37) |
| Meters rise in 2100 relative to 1986–2005 | 0.43 (0.28–0.6) | 0.52 (0.35–0.70) | 0.54 (0.37–0.72) | 0.73 (0.53–0.97) |

Note: Values are medians and likely range (data extracted from Table 13.5 of *Climate Change 2013: The Physical Science Basis; Contribution of Working Group I to the Fifth Assessment Report of the Intergovernmental Panel on Climate Change*, ed. T. F. Stocker, D. Qin, G-K. Plattner, et al. [New York: Cambridge University Press, 2014]). Numerical values for each RCP scenario are for total RF in year 2100 due to all greenhouse gases. See text for additional details.

substances. Unfortunately, many of these, along with the original compounds, are potent, long-lived greenhouse gases. According to the IPCC
Fifth Assessment Report, as a group, chlorofluorocarbons and related
compounds (hydrochlorofluorocarbons, hydrofluorocarbons, perfluorinated carbon compounds, other chlorinated and brominated chemicals,
and sulfur hexafluoride) rank third among the long-lived greenhouse
gases in terms of their total effect on global warming. The best estimate
for their RF is 0.18 W/m$^2$.

### Nitrous Oxide

Nitrous oxide ($N_2O$) is the fourth of the common, long-lived greenhouse
gases, with an estimated RF of 0.17 W/m$^2$. Its atmospheric lifetime is 121
years. As shown in table 2.1, its global warming potential is very high—
268 times greater than carbon dioxide at twenty years and 298 times
greater at one hundred years after it enters the atmosphere as shown in
figure 2.5. Due to decreases in the atmospheric Freon concentration (perhaps the most widely used halocarbon), as mandated by the Montreal
Protocol, $N_2O$ ranks third in terms of its contribution to global warming.
Nitrous oxide emissions arise directly or indirectly from multiple sources,
largely related to agriculture and food production. These sources account
for about 60 percent of the anthropogenic emissions of this gas.[23] Nitrous
oxide arises from synthetic fertilizers, animal wastes and their management, nitrogen leaching and runoff, human sewage, and other sources. As
with other greenhouse gases, the atmospheric concentration of $N_2O$ has
risen dramatically since the onset of the Industrial Revolution. Ice core
and other data indicate that current concentrations are higher than at any
other time during the last eight hundred thousand years.

### Other Factors Affecting the Earth's Energy Balance

In addition to the greenhouse gases discussed thus far, there are other
atmospheric constituents that affect the energy balance of the earth. In
general, these are short-lived compounds, such as volatile organic compounds (other than methane) and carbon monoxide, which cause positive
forcing, or a net energy gain. They yield ozone, methane, and $CO_2$ as the
result of atmospheric chemical reactions. Black carbon also contributes to
energy gain. Oxides of nitrogen, produced in boilers and internal combustion engines, prevent energy gain, as do mineral dust, sulfates, and nitrates.

The interaction between clouds and atmospheric aerosols (tiny droplets and particles dispersed in the atmosphere) prevents energy gain (negative forcing) and has blunted the effects of industrial age greenhouse gas emissions. This effect is difficult to quantify and is responsible for a substantial amount of the uncertainty in the estimate of total RF. Finally, changes in land use have affected the earth's *albedo*, or the tendency of the earth to reflect solar energy, and contributes a small negative effect to the earth's energy balance. These factors are all discussed in detail in the IPCC Fifth Assessment Report discussion of the physical science basis for climate change.[24]

## The Climate of the Future

The IPCC Fifth Assessment Summary for Policy Makers begins its section on the climate of the future with this unequivocal statement: "Continued emissions of greenhouse gases will cause further warming and changes in all components of the climate system." Because we have not yet experienced the future, climatologists must rely on mathematical representations of the behavior of the earth's systems, or models, to predict future behavior. This strategy is used widely in virtually every branch of science, ranging from predictions of the behavior of subatomic particles to the interaction of neural systems in the brain. When a modeler creates a climate model, he or she divides the earth into small segments. Each segment has multiple compartments, such as the land, ice, oceans, and atmosphere. Each compartment contains components of interest, such as $CO_2$, water, and heat. Modelers want to predict the future behavior of these components. Equations based on the laws of physics and chemistry describe the intercompartmental movement of the model's components, (e.g., $CO_2$). Additional equations describe the movement of compartmental components (e.g., $CO_2$) between compartments in adjacent segments (e.g., the back-and-forth movement of $CO_2$ between the atmospheric compartments of two adjacent segments). Models include predicted emissions into a compartment (e.g., $CO_2$ emissions by power plants), a variable that can be defined by the modeler, as well as information about how a greenhouse gas behaves in earth systems, such as the simplified carbon budget shown in figure 2.3. In a final step, the model is solved to yield a desired result, such as the temperature of the atmosphere or the ocean's surface. Model complexity and the requirement for enormous computing power increase as the number of earth segments increases and as more variables and their behaviors are included in the model.

Understandably, scientists devote a huge amount of effort toward validating models, and models of the climate are no exception. In general, model validation depends on using actual measurements of past physical conditions, such as temperature, and $CO_2$ emissions to predict a future, but known, climate. For example, a modeler may take known data from 1900 to predict the present temperature. Since both of these elements are known, the behavior of the model can be evaluated. If the model uses past data and makes an accurate "prediction" of the present condition, the model gains validity. If the model is inaccurate, it is refined by using more accurate data and an improved understanding of the climate system until it works properly. In this way, modelers have made extensive use of direct and indirect measurements of temperature, greenhouse gas concentrations, and ocean temperatures from the past to "predict" climate conditions of the less distant past or present. As scientists learn more about the behavior of the earth's systems and as more data become available, modelers become more experienced, and their predicted results become better and more reliable.

Because climate models are complex and virtually all elements include some uncertainty (a universal feature of all scientific observations), the confidence limits associated with long-term predictions grow as the model attempts to make predictions farther into the future. The presence of uncertainty does not mean that the model is invalid. What it *does* mean is that a scientist has been realistic in his or her ability to make future predictions. Restated, the uncertainties associated with predictions of the climate of the future do *not* mean that global warming will not occur. It will.

One of the most powerful and important models was developed by NOAA's Geophysical Fluid Dynamics Laboratory. This model includes interactions between the atmosphere and the oceans and is one of the most sophisticated approaches in current use. Many of the conclusions presented in the IPCC's most recent assessment are based on this model and the use of four representative concentration pathways (RCPs) that specify greenhouse gas concentrations along with other variables and their effects at various times. The results of applying four RCPs are shown in figure 2.6, which shows future surface temperatures, and table 2.3,

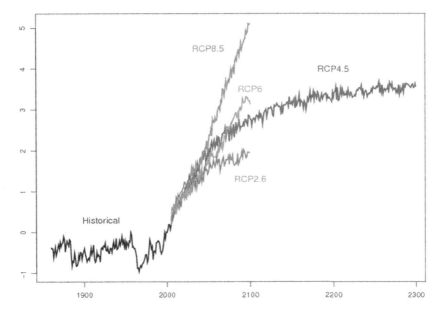

**Figure 2.6**

Representative concentration pathways. Global mean annual surface temperature changes (in degrees Celsius) simulated by the Geophysical Fluid Dynamics Laboratory at the National Oceanographic and Atmospheric Administration. Historical conditions (1860–2005) and four projected future RCP scenarios are shown. Named for the approximate RF in year 2100, the RCP scenarios include a low-forcing scenario (RCP2.6), two moderate-forcing stabilization scenarios (RCP4.5 and RCP6), and a high-forcing scenario (RCP8.5). As a work product of the federal government, this figure is not copyrighted and is in the public domain. Adapted from a color figure in Wikimedia Commons from data of D. P. Van Vuuren, J. Edmonds, M. Kainuma, et al., "The Representative Concentration Pathways: An Overview," *Climatic Change* 109 (2011): 5–31.

which portrays future sea levels predicted by the model scenarios. Each concentration pathway specifies the summed concentrations and effects of all greenhouse gases between the present and the year 2100 in terms of energy gain by the planet or RF (measured in W/m$^2$). Thus, RCP2.6 is a concentration pathway that yields an RF of 2.6 in the year 2100. This optimistic scenario includes a temperature peak partway through the century, with subsequent reductions in the temperature and in greenhouse gas concentrations due to efforts to control emissions. Temperatures stabilize after that year. The RPC4.5 and RPC6.0 pathways depict intermediate effects, whereas RPC8.5 depicts a business-as-usual scenario that predicts $CO_2$ concentrations will rise to about 950 ppm, just over twice the present concentration, by 2100. At that time, the model predicts a temperature increase of about 5°C with no end in sight and oceans that are 0.73 meters higher than they are now and rising at a rate that may be on the order of one-half inch per year.

### The Trajectory toward the Future

The overwhelming majority of climate scientists agree: the climate is warming, human activities are responsible, and, if we fail to act, there will be adverse effects that virtually all of us will feel. *Overwhelming majority* may be an understatement. The noted historian of science Naomi Oreskes analyzed the abstracts from 928 papers published in peer-reviewed journals to determine whether the authors endorsed the IPCC consensus position. "Remarkably," she wrote, "none of the papers disagreed with the consensus position."[25]

The RPCs that are the heart of the IPCC report outline four options. We are at a point where it is still possible to choose one. Yogi Berra, a former star baseball catcher for the New York Yankees, is famously said to have offered this advice: "When you come to a fork in the road, take it." The IPCC has presented us with four choices as depicted in the representative concentration pathways. We will take one of them, whether by choice or by default. It remains to be seen which one we will choose and where it will take us.

# 3

## Heat and Severe Weather

Things are not as simple as they seem.
—French public health official at the onset of the 2003 heat wave

### Heat Waves and Illnesses

The 2003 European heat wave began quietly, with news stories that were often humorous. A Danish taxi driver wore a skirt to work. He did this because his employer would not let him wear shorts. An Italian candy maker suspended delivery of chocolate eggs so they would not melt. German garbage collectors began their workday an hour and a half earlier than usual to make removal of "fast-rotting rubbish" more tolerable.

Soon, reports of an unusual number of deaths began to trickle in. Five people older than eighty-nine years of age died in a single weekend at a retirement home near Paris. Fifty Parisians were reported to have died from heat-related illness in just one week. The true magnitude of the problem started to become apparent when Les Pompes Funèbres Générales, France's largest group of undertakers, reported handling 50 percent more bodies than usual in a week. There were so many deaths that they had to rent refrigerated trucks to handle the unexpectedly large number of bodies. In what became a colossal understatement, a representative from the French health ministry was quoted as saying, "Things are not as simple as they seem." The final death toll throughout Europe attributed to the 2003 heat wave is thought to be around seventy thousand.[1]

This was not an isolated event. In August 2010, the Weather Underground reported on the combination of wildfire-fueled air pollution and heat that affected the Moscow area. The temperatures in the region exceeded 30°C for twenty-seven days in a row.[2] The Weather Underground quoted the head of Russia's weather service as saying, "Our ancestors haven't observed or registered a heat like that within 1,000 years." The toll from that event was expected to be at least fifteen thousand. The extreme heat combined with air pollution levels that were two to three times "maximum safe levels," and carbon monoxide levels soared to 6.5 times the so-called safe level.

## Heat Illnesses

Kory Stringer seemed unlikely to be susceptible to a heat-related illness.[3] This 335-pound, six-foot-four all-star offensive tackle for the Minnesota Vikings was twenty-seven years old when he took to the field for the last time on a hot summer day in 2001. On the day before he was stricken, he complained of exhaustion during practice. He was carted off the field, but vowed to return the next day. He did. On his final day, the temperature was around 90°F. The heat index, a measure of how hot it feels that is based on the humidity and actual temperature, was 110. The players were in full uniform. During the 2.5-hour practice, he vomited three times and left the field for an air-conditioned space. After complaining of dizziness, he was taken to a hospital, where his temperature was 110°F. He was unconscious until the time of death early the next morning.

Multiple heat-related medical conditions exist. These range in severity from heat rash to heat stroke, a life-threatening medical emergency and the cause of Kory Stringer's death. Certain medications, such as those needed to treat high blood pressure, some mental conditions, heart disease, and the extremes of age, all increase the risk of developing a heat-related illness.

*Heat rash*, also known as *prickly heat* or *miliaria*, occurs when the pores of sweat glands become blocked. This traps perspiration under the skin and causes itching and a raised, punctate, red rash. Heat rash is usually self-limited and rarely requires medical attention. Washing with cool water and general cooling measures are usually sufficient treatment. This condition is most common in babies and young children, but adults may also be affected. Heat rash becomes less common with aging, when skin changes and the number of sweat glands is reduced.

*Heat cramps* are painful spasms of muscles that may occur during exercise in hot weather. The calf muscles are commonly affected, but any muscle group involved in exercise may be involved. Treatment consists of cooling, rest, stretching, and massage of the affected muscles. These measures generally suffice to relieve symptoms. Sport drinks or juice may help, as long as there are no medical contraindications to their use. Although rare, very severe, prolonged cramps may be associated with damage to muscles that releases a muscle protein (myoglobin) into the blood. High concentrations of myoglobin in the blood may tint the urine, making it the color of strong tea, and may impair kidney function.

Under normal conditions, the body has reflex-like mechanisms that are designed to keep the body temperature within narrow limits. In response to heat and/or exercise, the blood vessels in the skin dilate and sweat glands produce more perspiration. The evaporation of water has a cooling effect. When this cooling mechanism begins to fail, patients may develop heat exhaustion. Symptoms of this condition typically include a cool, moist skin associated with heavy sweating. Unless the lost fluid is replaced, patients will become dehydrated. This is likely to lead to additional symptoms, including dizziness or faintness, weakness, or excessive fatigue. At that stage, patients may have a weak, rapid pulse and low blood pressure, particularly when standing up after sitting or lying down. Cramps, headache, and nausea may also be present. It is imperative to move the affected individual to a cool location and for that individual to stop all activity and rest. Replacement of the lost fluid by administering cool water or a sport drink is important. Persistent or worsening symptoms require urgent medical attention to prevent heat stroke.

*Heat stroke* is the most severe of the heat-related illnesses. It is a medical emergency and suspected victims should be transported to an emergency room in an ambulance. Heat stroke occurs when the defenses against a rise in body temperature are overwhelmed. Heat stroke is diagnosed when the body temperature reaches 104°F (40°C) or higher. The skin may be either hot and dry or moist. Profuse sweating does not reliably exclude heat stroke. Patients are frequently nauseous and may vomit, contributing to the loss of body fluids. The skin is usually red, as blood vessels dilate maximally. Respirations and heart rate may be rapid. Headache and confusion may progress to loss of consciousness. Emergency treatment is mandatory and should consist of moving out of the direct sun and removing excess clothing while waiting for emergency first responders. If possible, ice packs and wet towels should be applied. Fluids should be given by mouth unless the affected person is unable to swallow due to an impairment of consciousness. Failure to act promptly can lead to coma, irreversible damage to the brain, or death.

### The Epidemiology of Heat Illnesses

The US National Weather Service began to tabulate deaths due to severe weather in 1940 (www.nws.noaa.gov/om/hazstats.shtml, accessed April 22, 2014), at which time lightning was the leading cause of weather-related deaths. Lightning claimed 9,325 lives in the interval between 1940 and 2012. Like most other deaths due to weather, lightning deaths vary substantially from year to year. In order to smooth out the peaks and valleys between years, a ten-year moving average is a better descriptor. Moving averages are calculated for a year by averaging the data from the adjacent ten years. As years advance, the last year is dropped and the next year in the series is included. Using this method for reporting, the moving average for deaths due to lightning is smoothed out to thirty-five deaths per year. Heat fatalities were added to the database for the first time in 1986. Between then and 2012, 3,727 deaths were recorded, with a ten-year average of 117 per year, making heat the leading cause of weather-related mortality in the United States. The annual heat-death toll peaked at 1,021 in 1995, due in part to the Chicago heat wave.

Somewhat surprisingly, the Global Burden of Disease 2010 project does not list heat as a cause of death or a risk factor. However, the World Health Organization estimates that by 2004 global warming had become responsible for over 140,000 excess deaths each year.[4] In the Working Group II's contribution to the IPCC Fifth Assessment Report, heat is predicted to have a leading impact on health if global temperatures increase by 1.5C°. There will be further, dramatic increases if global mean temperatures rise by 4.0C°.[5]

*Heat wave* is a term that is used frequently, often without much or any precision—and there are good reasons for this. The effects of an increase in temperature are often complicated by other variables, such as the nature of the exposed population, the relative humidity, whether there is evening cooling or not, and others. The IPCC defines a heat wave in its glossary as "a period of abnormally and uncomfortably hot weather."[6]

This definition poses many problems, including the definition of a *period*. How should one define *abnormal* or *uncomfortable*, and what is meant by *hot*? The World Meteorological Organization has a more precise definition: when the daily maximum temperature on five or more consecutive days exceeds the average by 5°C, compared to the period between 1961 and 1990. This strict definition makes it possible for climatologists and others to make precise measurements and draw statistically rigorous conclusions about trends. Although this definition is precise, it fails to include the effects of other variables that determine the impact of temperature on health.

The problems associated with defining a heat wave call to mind Justice Potter Stewart's opinion with regard to obscenity in the landmark Supreme Court ruling in *Jacobellis v. Ohio*: "I know it when I see it." Thus, it may well be that virtually any definition of heat wave will have shortcomings—but people who are living through one know it. The IPCC may have had a valid reason for publishing a definition that lacked precision.

**Anatomy of a Heat Wave**

Most heat waves come and go with little fanfare or documentation. The 1995 heat wave that affected Chicago, Illinois, is an exception.[7] Like other heat waves, it was the result of a period of sustained high pressure in the upper atmosphere coupled with an unusually high level of moisture in the ground-level air. The official start of the heat wave occurred on July 12, when the temperature at both Chicago O'Hare and Midway airports reached 97°F, and it ended on July 16, when temperatures finally fell to 93°F at both sites—a temperature that was more typical for that time of the year.[8] Although the National Weather Service issued warnings, the highs and lows for those days exceeded predictions. The forecasts were not well covered by the media, and therefore the public was generally unaware of the impending heat wave and the risks it faced. In the United States, it is unlikely that weather forecasters would fail to issue similar warnings today: weather is big business for local TV stations. However, in other parts of the world where weather forecasts are either poor or nonexistent, populations at risk would have little if any opportunity to prepare let alone take action to prevent heat-related problems.

The number of heat-related deaths varied somewhat among reports. The National Oceanographic and Atmospheric Administration's report documented 495 heat-related deaths in Chicago between July 11 and July 27. In Milwaukee, Wisconsin, located about ninety miles (about 150 km) north of Chicago, there were eighty-five heat-related deaths during the same time interval.

The healthcare system in Chicago was stressed by the heat wave. During the peak week of the heat wave, hospital admissions were 11 percent higher than normal.[8] Admissions of the elderly, those over sixty-five years of age, were up 35 percent. They usually required treatment for dehydration, heat stroke, and heat exhaustion, and these diagnoses accounted for most of the increase. Chronic medical conditions were additional risk factors for hospitalization. These included cardiovascular disease (up 23 percent), diabetes (up 30 percent), renal disease (up 52 percent), and nervous system disorders (up 20 percent).

Of course, people die every day—so how can we attribute a death to a given heat wave? One study did just this. Investigators reexamined the toll exacted by the Chicago heat wave by comparing deaths during the heat wave interval with fifty-day periods from prior years that were centered on the day of the temperature peak.[10] They estimated that there were 692 excess deaths between June 21 and August 10. Deaths peaked on July 15, two days after the temperature peak, when 439 deaths occurred. The week of July 14 was the deadliest interval, when an average of 241 people died each day. There were 1,686 deaths that week: only 4.7 percent of these were reported to be due to heat alone, 28.1 percent included heat as a contributing factor, and 93.7 percent involved some form of underlying cardiovascular disease. Only a quarter of the deaths were considered somewhat premature. That is, these people would have died soon anyway, but the date of their demise came sooner than it might have. Technically, this is known as *mortality displacement*, or, more morbidly, as a *harvesting effect*. The relative risk for all-cause deaths on June 13, 1995, was 1.74; that is, Chicagoans were 1.74 times more likely to die during the heat wave than during a comparable year when temperatures were normal. African Americans were selectively at risk, but the mortality displacement in this population was lower.

A year after the disaster, a cohort of 339 friends or relatives of those who succumbed was interviewed. They were chosen if the person who died was older than twenty-four and if the death certificate specified heat or cardiovascular disease with or without heat as a contributing cause of death.[11] An identical number of age- and neighborhood-matched individuals served as a control population. Those with known medical problems were at the highest risk. In this group of unhealthy individuals, those who had been visited by nurses—a marker for poor health—had the highest risk of death. This was followed, in order, by those confined to bed, those who were unable to care for themselves, and those who had mental problems, a heart condition, or a lung condition. Risks were lowered if individuals had been contacted by a city worker during the heat wave. Taken as a group, the greatest risk factors for dying were being bed-bound or living alone. Availability of air-conditioned places and access to transportation were protective factors.

There are important conclusions to be drawn from these retrospective analyses: it is possible to mitigate the effects of heat waves. Improvements in weather forecasting are critical. Heat waves are like any other natural disaster; preparation and timely action focused on those with the highest risk will save lives. For more, see the section of chapter 10 addressing adaptation to heat.

Chicagoans were not the only ones affected by heat. In an analysis of the data collected by the largest emergency room data system for the years 2009 and 2010, there were approximately 8,251 emergency room visits for heat stroke in the United States.[12] This yields an annual incidence of 1.34 emergency room visits for this disorder for every 100,000 Americans. The highest incidences occurred among males (1.99 visits per 100,000) and those over eighty, who had about 4.45 visits per 100,000. Three and a half percent of those taken to emergency rooms for heat stroke died. As expected, heat stroke was the most common during the summer months of June, July, and August and more common in the South (1.61 visits per 100,000).

Although the National Weather Service's data concerning weather-related deaths begins to identify heat as a critical health-related risk factor, the Chicago experience and now the broader US experience show that this statistic fails to capture the importance of heat-related illness in the United States. The problem is certain to be worse elsewhere where it is hot more often, the public health infrastructure is either nonexistent or poorly prepared to cope with heat waves, and adaptive measures, such as air conditioning, are minimal or nonexistent. These problems undoubtedly contributed to the more than 2,500 deaths reported in India in June 2015.[13]

**Our Hotter Future**

Under virtually every imaginable scenario, global and US surface temperatures will be hotter in the future than in the historical past. This was shown graphically in figure 2.6. This figure portrays likely temperatures for each of the four representative concentration pathways (RCPs) that are at the heart of the IPCC's Fifth Assessment Report. As seen in the figure, there is not much difference in the projected temperatures by mid-century. However, with the high-emissions scenario, RCP8.5, surface temperatures increase sharply thereafter. This means that there will be more hot days and more instances of heat-related illness in this version of the future. In their analysis of these data, the authors of the American Climate Prospectus predict that Americans are likely to endure between two and three times as many days with temperatures in excess of 95°F compared to the years between 1980 and 2001 under RCP8.5.[14] This is the IPCC scenario that predicts a business-as-usual future, with little in the way of effective measures to limit greenhouse gas emissions. This scenario predicts that many Americans will experience between 46 and 96 days per year when temperatures exceed the 95°F threshold. The authors of American Climate Prospectus combined their results with predictions from the Third National Climate Assessment and predicted that residents of the Southeast are likely to experience between 56 and 123 days, or almost one-third of the year, when temperatures exceed 95°F. Between 1981 and 2010, that number was nine. This huge additional heat burden is virtually certain to lead to additional morbidity from heat and other aspects of daily life that are dependent on temperature and on agricultural productivity.

The temperature of 95°F is not chosen arbitrarily for scrutiny. When the relative humidity is 100 percent, this is the maximum temperature at which a normal, resting, well-ventilated individual can maintain a normal body temperature by the evaporation of sweat. At higher temperatures, the humidity must be lowered or the individual must move to a cooler spot. Failure to act increases the risk for heat exhaustion or heat stroke, particularly if one engages in even relatively mild exercise. Threats posed by the combination of high temperatures and humidity are less likely to result in an increase in heat-related illnesses in the already-hot southern portion of the United States because of the ready access to air conditioning.[15]

Two contemporary studies differ in their predictions of who will be the most susceptible to rising temperatures. One group predicts that temperatures above 80°F will have the highest impact on those between the ages of one and forty-four years of age.[16] This appears to be true in spite of an 80 percent reduction in mortality associated with heat after 1960 compared to heat-related mortality before that date. The protective effect after 1960 was due to air conditioning, in spite of the effects of heat on individuals with cardiovascular or respiratory diseases. Another group, using different methods, found that children less than a year old are the most susceptible to heat.[17] Both groups agree that mortality rates are lowest when average daily temperatures are between 50°F and 59°F. Below that range, mortality due to respiratory diseases rises in association with colder temperatures. Above 90°F, mortality rises at a rate of 0.08 percent per degree. In an analysis of the relationship between heat and hospitalization rates in California, study authors found a much higher rate.[18] They examined temperature and data for over two hundred thousand deaths, and found that a 10°F increase in the temperature was associated with a 213 percent increase in mortality.

The authors of the American Climate Prospectus conclude that heat-related mortality rates are not likely to change prior to the middle of this century, in accord with the various RCP scenarios that predict temperature changes in the future. This situation changes dramatically by the end of the century. The highest emissions scenario leads them to conclude that there will be between 3.7 and 21 deaths per one hundred thousand people. In a somewhat less likely conclusion, they predict that there will be a one in twenty chance that death rates will be higher than thirty-six per one hundred thousand. Table 3.1 presents a more complete picture of their results for different age groups for the RCP2.6 and RCP8.5 scenarios.

A more global approach to evaluating heat-related mortality has been taken by a group of Japanese investigators.[19] Using complex modeling techniques, they identified an optimum temperature at which the fewest deaths occurred in the Tokyo region, where year-round temperatures average about 28°C. From this baseline, they used temperature and

Table 3.1

Future climate change impacts on US mortality rates in the 2080 to 2099 time interval for the RCP2.6 and RCP8.5 scenarios

| Age group | RCP2.6 | RCP8.5 |
|---|---|---|
| Less than one year of age | -1.1 to 2.5 | 3.2 to 17 |
| 1–44 years of age | 0.2 to 1.5 | 3.1 to 7.6 |
| 45–64 years of age | -1.3 to 2 | 2.8 to 14 |
| Older than 65 years of age | -25 to 17 | -21 to 90 |

*Note*: The likely change range, in deaths per one hundred thousand, represents the 17–83 percent confidence limits. Extracted from T. Houser, R. Kopp, S. M. Hsiang, et al., *American Climate Prospectus: Economic Risks in the United States* (New York: Rhodium Group, LLC, 2014).

other variables such as population size and degrees of adaptation to estimate the number of expected excess deaths due to heat in various World Health Organization regions in 2030 and 2050. In the North American high-income segment, they projected a population size of 401 million in 2030, with a baseline mortality of about 2,400 deaths. Assuming 0, 50, and 100 percent adaptation, they estimated that there would be around 7,400, 4,700, or 2,700 excess deaths, respectively, due to heat. By 2050, the population projection was expected to be around 447 million, with around 15,400, 7,900, or 3,200 excess heat-related deaths, respectively, depending on the level of adaptation. These results buttress the assertions made by the IPCC in terms of the value of adaptive measures designed to reduce the health effects of heat.[20] Table 3.2 shows additional data.

## Heat and Healthcare Delivery

It is impossible to conclude that climate change is the cause of any given heat wave. However, it is possible to estimate the probability of recurrent episodes of hot weather. One group of climate scientists has done just that. They concluded that human activity has more than doubled the probability of experiencing an even worse heat wave than the one that gripped Europe in 2003.[21] Another group reached a similar conclusion.[22] Working from a baseline taken between 1999 and 2008, this group concluded that the probability was greater than 95 percent that a 2003-like

Table 3.2

Excess deaths due to heat for various World Health Organization regions

| Region | Est. population* | 1960– 1990 deaths* | 2050 deaths, no mitigation | 2050 deaths, 50 percent mitigation* |
|---|---|---|---|---|
| East Asia | 1,332,115 | 7,050 | 36,739 | 18,612 |
| Southern Asia | 2,284,943 | 17,489 | 65,562 | 37,524 |
| Southeastern Asia | 777,181 | 2,487 | 19,662 | 8,371 |
| Western Europe | 446,713 | 4,425 | 13,367 | 8,334 |
| Central Latin America | 318,626 | 1,226 | 7,364 | 3,363 |
| Tropical Latin America | 233,166 | 563 | 6,546 | 2,545 |
| North America High Income | 446,749 | 2,953 | 15,441 | 7,877 |
| North Africa/Middle East | 681,137 | 3,209 | 15,331 | 7,940 |
| East Sub-Saharan Africa | 848,878 | 1,679 | 14,743 | 6,060 |
| Central Sub-Saharan Africa | 212,327 | 438 | 4,007 | 1,645 |
| West Sub-Saharan Africa | 809,872 | 1,578 | 9,468 | 4,465 |

*Note:* Only those regions with projected populations of over two hundred million are shown. Data are from Y. Honda, M. Kondo, G. McGregor, et al., "Heat-Related Mortality Risk Model for Climate Change Impact Projection," *Environmental Health and Preventative Medicine* 19, no. 1 (2014): 56–63.
*Numbers are in thousands.

heat wave had "more than doubled under the influence of human activity in spring and autumn, while for summer it is extremely likely that the probability has at least quadrupled."[23] These and other findings have profound implications for those charged with planning for the future and how to mitigate the effects of climate change.

Not everyone is equally at risk for the development of a severe illness or death during heat waves. The Chicago experience showed that living conditions were important in determining the risk of death.[24] Markers for social vulnerability and poverty, such as absence of air conditioning, confinement to bed, and the need for Meals on Wheels, were markers for increased risk. Listening to the radio and reading newspapers were found to be associated with an increased knowledge of the health risks associated with the heat wave.

Fewer people die during mild winters. This defers death for some, placing those who might have died during the winter at greater risk for hospitalization and death during an ensuing hot summer.[25] Although confinement to a healthcare facility might be expected to reduce risk, this is not always the case. A comprehensive evaluation of the effects of the 2003 heat wave on hospitals in the United Kingdom revealed a number of ways hospitalized patients were vulnerable.[26] Problems with the power grid, including failure of freezers and information technology equipment, are risk factors for those in the hospital. The study found that the use of portable air-conditioning equipment taxed already strained power supplies and led to power failures. Some laboratory equipment was not adequately heat resistant and failed due to the heat. Finally, the heat itself had unspecified adverse effects on the hospital staff and patients.

It is almost a "no-brainer" to conclude that warming due to climate change will lead to more severe effects on health. In one quantitative approach to predicting the effects of heat on health, two investigators from the London School of Tropical Medicine and Hygiene evaluated data from over one hundred US communities between the years 1887 and 2000 in order to determine whether prolonged spells of hot weather carried an extra risk for death beyond that associated with the usual risk for each day. The investigators concluded that there was a daily risk from high temperatures. They referred to this as the *main effect*. A so-called added effect of sustained high temperatures appeared after four days of hot temperatures. Although the added effect was important, it was outweighed by the main effect.[27]

**Mitigation of Heat Effects**

One of our challenges will be to develop and use effective measures designed to adapt our social systems and the built environment to higher temperatures. Adaptation works. Largely because of the spread of air conditioning, investigators have found that the mortality on hot days fell by around 80 percent in the years from 1960 to 2004 compared to the years between 1900 and 1959.[28] Before 1960, these authors presumed that there were around 3,600 deaths annually compared to six hundred premature deaths after 1960. This information has serious implications for the future in both the United States and the rest of the world as surface temperatures climb, regardless of emissions. A separate study estimates that residential electricity bills will increase by 15–30 percent to pay for the increase in air conditioning.

Risks associated with heat waves are like other risks; they are a function of the hazard, the vulnerability to the hazard, and the level of preparedness. The published experiences of others tell us what we need to do. Cities need to be designed and reengineered to resist the effects of heat and the buildup of trapped heat caused by the combined effects of hot weather, building design, and energy use, the so-called heat islands. Effective measures include planting trees and other vegetation, sometimes on the roofs of new or existing buildings. These measures reduce the need for air conditioning, trap and use storm water, and enhance the aesthetic value of property. Reflective, cool roofs produce similar benefits by reducing the need for air conditioning. Cooperation between city governmental agencies and nonprofit institutions, such as Visiting Nurse Association, Meals on Wheels, and others, can provide a mechanism to monitor the health of the individuals at the greatest risk: the sick and those who live alone who may be home- and bed-bound. The Chicago experience, like that of other municipalities, points out the need for cooling shelters for those without air conditioning. Hospitals and other healthcare institutions need disaster plans that include heat emergencies. Planning for heat emergencies will save lives. Additional mitigating factors are discussed in chapter 10.

## Extreme Weather Events

Aside from heat, most extreme weather events do not have specific health problems identified with them. Nevertheless, cyclones and hurricanes, floods due to intense precipitation or storm surges, tornadoes, and other weather events take a toll on lives. Climate change is likely to have effects on these extreme weather events.

Warming of the climate drives extreme weather events. As the earth's surface, atmosphere, and oceans warm, fundamental physical principles dictate that more water will evaporate from bodies of water and the soil. This increase in the water content of the atmosphere leads to storms and increases in precipitation, deaths and injuries, and property damage. According to the National Climatic Data Center, there have been 151 weather or climate disasters since 1980 for which costs were at least $1 billion (adjusted to 2013 dollars). The total estimated cost is over $1 trillion. Seven of these events took place in 2013.

### Extreme Precipitation

Heavy rains and flooding are big news. The following samples are culled from reports from May 16, 2014, a date chosen for no particular reason: "Record rain floods Triangle roads ... knock out power [as] a record 3.38 inches of rain," *Raleigh (NC) Newsobserver*; "Isolated storms, heavy rain pose flood risks Friday," *Baltimore Sun*; "Heavy rains flood streets, creeks, and cancel flights in North Texas," *Dallas Morning News*; "Serbia and Bosnia-Herzegovina have been hit by some of the worst flooding in each country's history," and "A plodding system that has left flooding in multiple Midwestern and Southern states continued to creep up the East Coast, bringing a serious flash flood threat and major flooding to areas of the Mid-Atlantic," according to the Weather Channel.

Rain is common in the Pacific Northwest, but the record book went out the window in the spring of 2014. Rainfall was 200 percent above normal for the forty-five days prior to March 22, 2014, when a landslide obliterated a small community four miles east of Oso, Washington.[29] Soil saturation had created an unstable condition that caused the collapse of a hill. The landslide that resulted engulfed an area of approximately one square mile. At least forty-one people died.

Although any one of these single events may be completely unrelated to climate change, as a group they are consistent with the pattern of increased rainfall observed over the past several decades while global surface temperatures have been increasing and with predictions for the future.

A detailed sixty-two-year study of the rainfall over the central portion of the United States—a region that includes Texas, Louisiana, and parts of Mississippi north through the Mississippi River valley to Minnesota, Wisconsin, Michigan, and Ohio—was published in 2012.[30] The investigators included just those sites that had rainfall-measuring devices that met strict, predefined criteria for accuracy. They found a significant redistribution of rainfall when they compared the 1948–1978 era to a thirty-one-year period that ended in 2009. There was an increase in the frequency

Table 3.3
Observed changes in heavy rainfall patterns in the United States, 1958–2012

| Region | Percent change in heaviest 1 percent |
|---|---|
| Northeast (ME, VT, NH, MA, NY, PA, NJ, DE, MD, WV) | 71 |
| South (VA, KY, TN, NC, SC, GA, AL, MS, FL, LA, AR) | 27 |
| Midwest (OH, MI, IN, IL, WI, MO, IA, MN) | 37 |
| Great Plains (MT, ND, SD, WY, NB, KS, OK, TX) | 16 |
| Pacific Northwest (WA, OR, ID) | 12 |
| Southwest (CA. NV, UT, AZ, NM, CO) | 5 |
| Alaska | 11 |
| Hawaii | -12 |
| Puerto Rico | 33 |

of days with "very heavy" rainfall exceeding three inches (76.2 mm) and "extreme precipitation events," defined as daily rainfall exceeding six inches (154.9 mm). In this region, there were sites with up to a 40 percent increase in the frequency of daily and multiple-day occurrences of these extreme precipitation periods. Tropical storms did not contribute to these results. The intensity of precipitation—that is, hourly totals—remained constant during the sixty-two years of the study period.

This regional trend is illustrative of the changes in rainfall patterns throughout the United States. Although the total amount of rainfall in the United States has increased by about 7 percent during the past one hundred years, the amount of rain falling during the heaviest downpours (the highest 1 percent) has increased by as much as 71 percent in the Northeastern part of the nation.[31] Table 3.3 shows regional change data.

## Hurricanes and Cyclones

Few natural events unleash more power than tropical hurricanes and cyclones. The web provides many values for the amount of energy they release. One of the most reliable calculations has been made by the NOAA Hurricane Research Laboratory.[32] The laboratory estimates that a mature hurricane releases about $1.5 \times 10^{12}$ watts per day, an amount roughly equal to about half of the electricity-generating capacity of the entire world! No wonder these storms often cause tremendous amounts of damage.

Because these storms have such a high potential for causing huge amounts of damage from their wind, rain, and storm surges, climatologists have struggled to predict the effects of climate change on their frequency, intensity, and the paths they are likely to follow. The absence of a longstanding accurate historical record is a significant problem. Aircraft were first used to monitor severe tropical storms in the 1940s. Although planes are still used, the data collected during these flights have been augmented by satellites first launched in the 1960s. The best data are from the North Atlantic, as monitoring is most intense over this region. These data have shown that there are periodic oscillations in hurricane activity in this region, known as the Atlantic multidecadal oscillation. These oscillations are thought to be due to normal factors, which form a starting point for modeling studies designed to predict future activity.

Tropical climates control the activity of hurricanes.[33] Therefore, changes in tropical climates, whether they are due to human or natural activity, can be expected to have important effects on hurricane activity. Volcanic eruptions act to cool the tropics as a result of the injection of dust (particulates) and sulfur dioxide into the stratosphere. Greenhouse gas emissions have the opposite effect. As the result of the Clean Air Act, particulate emissions by US sources have declined significantly, thereby diminishing their cooling effect. This is particularly true for the sulfate-containing particles that originate primarily from burning coal.[34] Dust from the Sahara Desert, due to the combination of natural and human activity, also has an effect on the temperature of the tropical North Atlantic. The sum of all of these agents acting in opposite directions helps determine the temperature of the sea surface and hence the probability and strength of hurricanes.

When these factors are incorporated into climate models, it seems likely that the number (or frequency) of tropical cyclones, including hurricanes, will not change much in the future. That is the good news. The bad news is that the models predict that of the storms that do form, more of them will be in the category 4 and 5 range, as defined by the Saffir-Simpson scale.[35] According to this scale, category 4 storms have winds between 130 and 156 mph (209–251 km/h), and category 5 storms have winds greater than or equal to 157 mph (252 km/h).

Wikipedia is a rich source of varied information about Hurricane Sandy, also referred to as Superstorm Sandy. Sandy was only a category 3 storm at its peak when it made landfall in Cuba. Its winds diminished to category 2 (96–110 mph, or 154–177 km/h) when it reached the North Atlantic. However, it was huge, with winds reaching over a diameter of 1,100 miles. Because of its size, the configuration of the New York Harbor region, and the fact that the storm surge coincided with the high tide in New York, it caused about 286 deaths in the United States and $68 billion in property damage. Healthcare institutions were poorly prepared, and many were damaged severely. Bellevue, New York University's Langone Medical Center, and Coney Island Hospital had to be evacuated after multiple critical aspects of their infrastructure failed.

Katrina was a category 3 hurricane when it made landfall in Louisiana.[36] She was one of the costliest storms ever, causing over $110 billion in damages and claiming over 1,800 lives. For a chilling account of the effects of Hurricane Katrina on Memorial Hospital in New Orleans, read the Pulitzer Prize–winning *Five Days at Memorial: Life and Death in a Storm-Ravaged Hospital*, by Dr. Sheri Fink.[37]

## Severe Thunderstorms and Tornadoes

Few weather events are more dramatic than tornadoes. Although forecasting has improved substantially, they often strike suddenly with little warning. Severe property damage and loss of life are not rare when tornadoes strike population centers.

Severe thunderstorms may be accompanied by strong winds, hail, torrential rain, and, on occasion, tornadoes. Severe thunderstorms that rotate are known as *supercells* and are the most likely to spawn tornadoes. Thunderstorms are caused by warm air that rises rapidly in association with large differences in surface winds and winds at about 6 km above the earth. The technical term for the tendency for air to rise is related to the amount of energy available when a segment of the atmosphere is lifted a defined distance in the atmosphere: *convective available potential energy* (CAPE). Warm water in the Gulf of Mexico, typical wind patterns, and the presence of the Rocky Mountains interact to make the Great Plains and the Eastern part of the United States particularly susceptible to the development of these events.

Although the ability to make short-term predictions and issue warnings about severe thunderstorms has improved dramatically, very long-term predictions that extend to the end of the century are much more tenuous. One problem centers on the absence of high-quality data from the past. Changing definitions and the absence of data-recording sites have made it difficult to characterize the frequency of these storms in the recent past.

Several recent reports illustrate the difficulty in predicting the effects of climate change on thunderstorm activity. One such analysis concludes that CAPE will increase as the climate warms, a change that would be likely to increase storm frequency.[38] However, the author of this analysis, H. E. Brooks, concludes that this effect will be countered to a degree by changes in wind shear that are likely to occur simultaneously. Brooks concludes that the accuracy of long-range predictions is likely to be problematic. A group from Stanford and Purdue examined the lack of synchrony between CAPE and wind shear and determined that the warming climate will lead to more days when the relationship between these variables will favor the formation of severe thunderstorms.[39] They conclude that severe thunderstorms will indeed become more common unless greenhouse gas effects are lessened.

## The Trajectory toward the Future

The earth's temperature increases as climate change becomes more pronounced, and this will have multiple effects. Heat waves will claim more lives, particularly in developing nations and among the socially disadvantaged, due to heat-related illnesses such as heat stroke and heat exhaustion. Heat will evaporate more water from lakes, rivers, and oceans, causing increases in precipitation in some areas and droughts in others. More intense hurricanes are likely. It is probable that there will be more tornadoes and severe thunderstorms. Humans will not do well in the heat, and neither will the earth's ecosystems.

# 4

# Infectious Diseases

The great progress that has been achieved could be undone in some places in a single transmission season.
—Dr. Margaret Chan, WHO Director-General, *World Malaria Report 2013*

It is not difficult to see why heat-related illnesses will become more common and pervasive on a warming planet. The evidence that climate change will increase the risk of certain infectious diseases is more complex but just as compelling.[1] Many diseases that are of greatest concern have multiple facets that are susceptible to the effects of climate change. These include the physical properties of ecosystems that affect diseases and disease vectors, such as temperature and moisture, and the numbers and types of species in the system, including species extinction. Human activity has already had and will continue to have an enormous impact on ecosystems and the species that depend on them. Understanding how climate change affects each of these factors is critical to developing and implementing efforts to mitigate and adapt to climate change and to control diseases such as malaria, dengue, West Nile fever, and others.

Changes in the locations where insect vectors live is perhaps the most obvious factor affecting vector-borne diseases. For example, changes in precipitation impact the breeding cycle of mosquitoes, such as the females in the genus *Anopheles*, carriers of malaria, and *Aedes aegypti*, the principal mosquito vector for dengue fever. Changes in temperature may affect the reproductive cycle of malaria parasites and the mosquito vector.

The need to integrate all of these variables has given rise to a new scientific specialty known as *disease ecology*.[2] Disease ecologists study the relationships between climate and its effects on the physiological state of the pathogen and the population of disease carriers, or *vectors*, including the balance of species in an ecological system. Disease ecologists are concerned increasingly about a phenomenon known as the *dilution effect*. When the diversity of intermediate host species increases (e.g., the number of mosquito species increases), there is a reduction in the exposure to disease(s) carried by a specific species. This is known as *dilution*. In other words, diluting the species with multiple species reduces risk. Risks increase when there is less diversity among species, as has been shown to be the case for Hanta virus exposure and Lyme disease. The risk of both diseases has been shown to rise as species diversity falls. Similarly, the risk of West Nile virus exposure rises as diversity among bird species falls.

Application of the principles of disease ecology is beginning to have important implications for public health. A case study of malaria in Botswana illustrates a successful application of the disease ecology approach.[3] Malaria is a major public health problem in Botswana in spite of the fact that it is a semiarid nation, and Botswana suffered a major malaria epidemic in 1996. The country's malaria control strategy was developed from a retrospective analysis of the seasonal variability of malaria and a multimodal approach to climate prediction based on sea-surface temperatures. Variations in the temperature of the ocean are, in turn, the result of predictable fluctuations in prolonged oceanic warming, known as *El Niño*, and the corresponding atmospheric pressure component, known as the *Southern Oscillation*. Together, these components are referred to as the *El Niño–Southern Oscillation* (ENSO).

From ENSO data, disease ecologists were able to make more accurate climate predictions for southern Africa. Improved climatological predictions in turn lead to better forecasts for malaria risk. Using these combined data, public health officials were better able to allocate resources needed to combat an impending malaria epidemic. This process involved judicious use of insecticides, administration of drugs to individuals during periods of high risk to prevent malaria (chemoprophylaxis), and the early detection and management of the disease in individual patients.

This ability to reference changes in disease prevalence to distant climatological phenomena, such as ENSO, is referred to as making a *teleconnection*—literally, connecting distant events. Other recent evaluations of teleconnections have led to the discovery of links between Rift Valley fever and periods of heavy rainfall and between the disease known as chikungunya and high temperatures coupled with drought.[4] Both of these diseases are carried by mosquitoes. In the case of Rift Valley fever, high rainfall amounts allow the vector to flourish. In contrast, chikungunya outbreaks are thought to be due to in-home water storage that is increasingly common during drought. The common denominator is an increase in the mosquito vectors, but the mechanisms that underlie the increases are quite different.

Contemporary strategies for controlling diseases carried by insects rely increasingly on combining multiple techniques known collectively as integrated vector management (IVM). The World Health Organization defines IVM as "a rational decision-making process to optimize the use of resources for vector control. IVM requires a management approach that improves the efficacy, cost effectiveness, ecological soundness and sustainability of vector control interventions with the available tools and resources."[5]

As implied by this definition, the first step in the process is to understand local factors that affect vector ecology and then select the control measures from a range of available options. These include environmental management (e.g., draining standing water), the use of biological controls (e.g., fish that eat larvae or bacteria that kill larvae, such as *Bacillus thuringiensis* [Bt]), chemical controls such as insecticides, and personal protective measures such as avoiding outdoor activity when mosquitoes feed, wearing long-sleeved shirts and long pants, and using insecticide-impregnated sleeping nets. Integrating these methods is designed to have a maximum effect on disease prevention with a minimum of risks to health and/or environmental effects of the control mechanisms.

### Arthropod-Borne Diseases

"Arbovirus" is an acronym for "arthropod-borne virus." These viruses are carried by arthropods, invertebrate animals with an external skeleton or cuticle (made of chitin with or without calcium carbonate) and with jointed appendages. Examples include insects, spiders, ticks, other arachnids, and crustaceans. (Try not to think of bugs when sitting down to a tasty lobster or crab dinner!) Insects are arthropods with three body parts, a head, a thorax, and an abdomen. They also have six legs, compound eyes, antennae, and often wings. Mosquitoes are the most important vectors for diseases caused by arboviruses. These diseases include dengue, St. Louis encephalitis, Zika, and West Nile fever. Ticks are also arthropods. However, they have two body parts and eight legs and do not have antennae or wings. Lyme disease is the most important tick-borne disease.

Malaria parasites are *protozoa*, single-celled organisms containing a nucleus and other organelles. This section includes malaria along with the arboviruses because malaria parasites are carried by mosquitoes. Thus, factors that affect mosquitoes affect the diseases caused by arboviruses and malaria.

### Dengue

*Dengue*, also known as *breakbone fever*, is a leading cause of death and disease in tropical and subtropical regions. A 2011 comprehensive review concluded that dengue's prevalence had increased by a factor of thirty during the preceding five decades and was now endemic in 112 nations.[6] Estimates conclude that in 1990 around 30 percent of the world's population had a greater than 50 percent chance of contracting the disease.[7] The authors of this study used predictions of the future climate and the ecology of its mosquito vector to estimate that by 2085 between five and six billion people will be at risk for this disease. Their model predicts that the probability of transmission will be greater than 0.2 for large portions of the south-central and southeast portions of the United States, with the probability rising to 0.9 or more in sections of the Texas gulf coast and Southern Florida. (Note: Probability is expressed as a number ranging from 0.0, impossibility, to 1.0, certainty.)

Currently, dengue is relatively rare in the United States. Most verified cases have been found among those who have returned to the United States after travel to countries where the disease is common. For example, a June 4, 2014, report in *Time* described twenty-four confirmed cases of dengue in Floridians who had traveled to regions where dengue is prevalent.[8] A more serious outbreak of dengue occurred in Hawaii in February 2016, leading Governor David Inge to declare a public health emergency.

Dengue is a viral disease. It is caused by RNA viruses in the genus *Flavivirus*. This is a particularly nasty genus; other members of the group cause yellow fever, West Nile fever, St. Louis and Japanese encephalitis, and others. Collectively, they are known as arboviruses because they are typically transmitted by arthropods. Several species of the mosquito genus *Aedes*, principally *A. egypti*, transmit the disease. These mosquitoes are well adapted to urban living and thrive in regions that are warm and humid—that is, tropical and subtropical regions at relatively low elevations. Although there is hope on the horizon that vaccines for the virus will be forthcoming, none are available at present. (In chapter 10, I discuss the progress toward developing a vaccine.)

Therefore, preventive measures designed to control mosquitoes through the techniques of integrated vector management are needed. These measures include eliminating standing water and preventing mosquito bites. As the planet warms and atmospheric water increases as a result, it is expected that conditions that favor increases in the ecological niche will promote the propagation and spread of *Aedes* mosquitoes. Thus, regions where dengue is rare or virtually nonexistent will shrink as the latitude and elevation boundaries that favor mosquito-breeding change.

There are four strains or *serotypes* of the dengue virus (DEN-1, -2, -3, and -4) that differ in their virulence, which creates the possibility of multiple infections in the same person by different strains of the virus. Dengue is usually asymptomatic or mild, and the patient may not associate the fever with the disease. However, some patients (fewer than 5 percent) develop low blood pressure and shock due to leakage of blood products from blood vessels. The most severe infections are associated with hemorrhage—hence the name *dengue hemorrhagic fever*. Unfortunately, contracting a mild case of dengue does not confer immunity on the host. To the contrary, reinfection with a virus of a different serotype may cause disease that is much more severe than the first time around. The reasons for this immunologically based phenomenon are complex and incompletely understood.

The hemorrhagic form of dengue is most common in infants and children but may occur in adults. Children with excellent nutrition are paradoxically more likely to develop severe manifestations of the disease than those who are undernourished. The protection associated with undernutrition may be due to the poor immunological response associated with protein-calorie deficiency.[9] This may be the only silver lining associated with the dark cloud of undernutrition.

### West Nile Fever

West Nile fever is caused by another of the mosquito-borne viruses in the *Flavivirus* genus. As noted previously, other members of this genus cause dengue and other diseases. Although rare cases were identified before the mid-1990s, West Nile fever became relatively common in Algeria after that time. The first US case was identified in 1999 in New York City. It spread rapidly across North America during the next five years.[10]

West Nile virus has several host reservoirs. It infects humans after the bite of a mosquito that carries the virus. These mosquitoes are primarily in the genus *Culex*, with regional variations in the species.[11] In the United States, there are approximately sixty different species in the *Culex* genus, but fewer than ten are thought to be the main vectors, which vary regionally. *Cx. pipiens* (the northern house mosquito) is responsible for more than half of the isolates in the Northeast, *Cx. quinquefasiciatus* (the southern house mosquito) predominates in the South, and *Cx. tarsalis* predominates west of the Mississippi River. The latter is the most efficient transmitter of the disease. Crows and robins act as reservoirs for the disease and are partially responsible for spreading the disease from one location to another.[12] Monitoring deaths of these birds has proven to be a valuable means of detecting spread of the disease.

Fears have arisen that globalism and the ease of air transportation will lead to rapid disease transmission. This was one theme of the 2011 film *Contagion*. In the film, Gwyneth Paltrow's character, Beth Emhoff, dies after flying home to the United States after contracting a disease in China. A global epidemic ensued, spread initially by Emhoff's fellow passengers on her flight. In the real world, travelers to the 2014 FIFA World Cup (soccer or football) were warned about dengue transmission when they traveled to Brazil, the nation with the largest number of cases of dengue. Brazil was gripped by an epidemic of the disease during the games. There are similar fears surrounding the 2016 Summer Olympics, scheduled for Brazil.

Although most cases of West Nile fever are unreported, because the symptoms are generally mild or unnoticed, the Centers for Disease Control reported that 2012 was the worst year for West Nile fever in the

United States. In that year, there were 5,674 confirmed cases and 286 deaths. Similar numbers were reported in 2002 to 2003. An investigation of the 2010 European outbreak led to the conclusion that elevated temperatures were largely responsible for the epidemic. Similar conclusions were drawn for the 2012 US outbreak.

Climate change seems virtually certain to affect the prevalence of West Nile fever. However, we cannot draw a straightforward link between temperature and the spread of the disease. A 2013 study of the effects of climate change illustrates this point.[13] Using the IPCC A2 climate change scenario, which predicates an approximately fivefold increase in carbon dioxide emissions by 2100, and the so-called Dynamic Mosquito Simulation Model, investigators concluded that the population of *Cx. quinquefasciatus* will not be homogeneous and will depend on variables such as temperature and precipitation. For example, in some parts of the country the temperature will be too high for mosquitoes to breed, and in others it may be too dry. In South Florida and the Texas Gulf Coast, the study authors predict bimodal disease peaks at about weeks 20–25 into the year, and then later at weeks 40–45. In the arid southwest, they predict a larger single peak at about week 40. Some of the projected regional differences are based on the knowledge that when it is too hot, *Culex* mosquitoes fail to reproduce in large numbers. Thus, latitude, altitude, temperature, and precipitation all affect these mosquitoes and therefore the spread and prevalence of the disease.

### Chikungunya

Chikungunya is caused by a virus in the *Alphavirus* genus. The name *chikungunya* is reportedly derived from the Makonde language, meaning that which bends up, a reference to the posture assumed by some victims. Its symptoms include headache, fever, a skin rash (petechial or maculopapular), and joint pain that typically last for two days but may persist for many days or even months. The mortality rate is around one in one thousand. The virus can be recovered from some patients months after the initial infection. It is spread by mosquitoes in the *Aedes* genus, usually *A. egypti*.

About fifty years after the initial description of the disease, an outbreak occurred in Italy that was linked to *A. albopictus*.[14] This change in vector is thought to be due to a mutation in the virus that enables *A.*

*albopictus* to carry the disease. This is worrisome, as the *albopictus* mosquito is a more aggressive biter and therefore may be a more efficient transmitter of the disease. A review indicates that chikungunya is quite widespread, with as many as 244,000 cases on the island of La Reunion, a French island east of Madagascar, and a million cases in India.[15] The high prevalence on this island nation was thought to be due to a change in the vector and importation of the disease by travelers. There is concern that chikungunya may pose a threat in the United States. Although there are no FDA-approved vaccines for the disease, promising trials have been reported.[16] The development of a safe and effective vaccine would be a major step forward in controlling this disease.

### Zika Disease

In January 2016, a widely publicized climate-linked threat to health emerged in the form of the Zika virus.[17] The Zika virus belongs to the *Flavivirus* genus, the same genus as the virus that causes dengue. Like dengue, it is spread by *Anopheles* mosquitoes. Zika virus infections are likely to become more common as climate change increases the range of the insect vector. Monitoring the prevalence of dengue and chikungunya is likely to foreshadow the spread of Zika virus disease. The symptoms of Zika virus disease are usually mild and similar to those of dengue. Zika virus infections are thought to have severe adverse effects on developing brains, however, causing microcephaly, or small head and brain size. The risk of microcephaly is likely to be greatest among pregnant women infected during the first trimester but fetal infection may occur at any time. Microcephaly causes severe, life-long mental retardation. As a result, the Centers for Disease Control and Prevention issued a travel advisory urging pregnant women to avoid travel to regions where Zika virus has been isolated. There are no specific treatments for the disease nor is there a vaccine. Avoiding mosquito bites is the only effective preventive measure.

### Lyme Disease

Lyme disease is caused by *Borrelia burgdorferi*, a corkscrew-like bacterium also known as a *spirochete*. The Centers for Disease Control and Prevention (CDC) website is a rich source of information about this disease.[18] It is the most common of the vector-borne diseases in the United

States and is on the rise. In 2004, there were 19,804 reported cases. In 2013, this number grew to 27,203 confirmed and 9,104 probable cases. Most reported cases are from the northeastern part of the country or the upper midwest. Data from 2013 show that the incidence was highest in Maine, with one hundred cases per one hundred thousand people, and lowest in southern states, many of which reported no cases that year. As suggested by the number of cases, surveillance data show clearly that Lyme disease is spreading as well as increasing in prevalence in states where it already exists.

The disease is contracted after being bitten by an infected tick. Three to ten days later, between 70 and 80 percent of infected persons develop a characteristic rash called *erythema migrans*, a red, target-shaped rash centered on the site of the bite. Blood tests, when done properly, are diagnostic for the condition. Treatment with oral antibiotics almost always is curative, if given at an appropriate time after the infection is contracted. Lyme disease may have other manifestations that occur days or weeks after the bite, including Bell's palsy (drooping of one side of the face due to involvement of the seventh cranial nerve), rashes appearing on other parts of the body, severe headaches due to infection and inflammation of the meninges (membranes that cover the brain and spinal cord), arthritis (especially involving large joints, such as the knees), or irregularities of the heartbeat that cause palpitations or occasionally dizziness.

Relatively simple steps to avoid ticks may prevent the disease. These include avoiding areas where ticks are prevalent (woods, grassy areas), application of repellents—especially those containing DEET (N, N-diethyl-m-toluamide) or permethrin—conducting a whole-body search for ticks, and bathing after a potential exposure. Ticks may also be present on pets or other objects, such as clothing. Ticks should be removed using fine-tipped tweezers. After removing the tick, apply alcohol or an iodine scrub to the affected area or wash carefully with ordinary soap and water. Public service announcements by the media may be an effective and inexpensive method to teach individuals how to deal with ticks and prevent Lyme disease.

In the northeastern, mid-Atlantic, and north-central parts of the United States, the disease is usually spread via the blacklegged or deer tick (*Ixodes scapularis*). On the Pacific Coast, the western black-legged tick is the vector (*Ixodes pacificus*). Unfortunately, these are tiny ticks, and the immature nymph stage of development typically is the culprit. The nymphs are about the size of a poppy seed. Adults that have not had their blood meal are about the size of a sesame seed. Most bites occur during summer months.

Several studies have shown that migrating birds carry the nymph form of *Ixodes scapularis* as unwelcome passengers.[19] The tick climbs aboard the birds as they search for food on the ground. After birds complete their migration, the ticks detach themselves, complete their transformation into adults, and then spread the disease to other species, including humans or some other intermediate hosts. In this way, these tiny creatures are able to move into new territories where conditions are favorable for proliferation of the tick and the spirochetes they may harbor.

Based on what is known about the conditions necessary for survival of the ticks, it seems highly likely that Lyme disease will continue its spread northward into Canada from locations in the United States. Modeling studies that project the northern boundaries where the temperature will remain warm enough during the winter to allow the ticks and thus the spirochetes to survive predict that Lyme disease may spread northward by as much as 1000 km by the 2080s.[20] It seems likely that warming that has already occurred has contributed to the spread of Lyme disease already observed.

## Malaria

Malaria has long been and remains one of the great scourges of mankind. Estimates of the actual number of individuals with malaria worldwide vary substantially. The Global Burden of Disease project reports an almost 20 percent increase in malaria mortality between 1990 and 2010, with the disease responsible for 1.17 million deaths in 2010.[21] The World Health Organization (WHO) paints a more optimistic picture, reporting a 45 percent reduction in all age groups and a 51 percent reduction in the 2000 to 2012 time interval.[22] In yet another time interval, global malaria deaths were estimated at 995,000 in 1980, rising to peak of 1,817,000 in 2004, then decreasing to 1,238,000 by 2010.[23] The deaths in 2010 in one report were estimated to be twice the number reported by the WHO for the next year (2011). In its 2010 Millennium Development Goals Progress Report, WHO touts a 20 percent reduction in childhood malaria deaths as progress toward its 2015 goal.

Some of the discrepancies in these numbers are the result of differences in the time intervals included in the various reports. In spite of the failure of experts to agree on the numbers, all of these reports emphasize severity of the problem and the need for increased financial support for malaria eradication efforts. Reduced financial support for malaria eradication efforts due to the recent worldwide financial crisis is thought to be the cause of setbacks in the control of the disease. The fragility of the attempts to control malaria was emphasized by Dr. Margaret Chan of the WHO in the *World Malaria Report 2013* quotation at the beginning of this chapter: "The great progress that has been achieved could be undone in some places in a single transmission season."

References to malaria date back about four thousand years. It was known to Hippocrates and is thought to be responsible for the decimation of several Greek city states and rural areas during the age of Pericles (461–429 BCE). There are references to the disease in Sanskrit and in Chinese writings. The Romans linked the disease with swamps. The properties of the Qinghao plant were described in Chinese manuscripts from around 340 CE. Derivatives of the active ingredient of this plant, known as *artemisinins*, are still a mainstay in the treatment of malaria. From these writings, we can conclude that malaria has been a worldwide problem for a great many years.

The Nobel Committee has recognized the importance of malaria research with the award of four prizes. The first went to Ronald Ross in 1902 for his description of malaria parasites in the gastrointestinal tract of mosquitoes and the elucidation of the complete life cycle of the parasites. Charles Louis Alphonse Laveran was awarded the 1907 prize for much earlier work that described malaria parasites in the blood of a malaria patient. Camillo Golgi found that the fevers that are characteristic of the disease coincided with the rupture of parasite-laden red blood cells; he was awarded the 1906 prize. In the summer of 2015, the prize was awarded to Youyou Tu for her development of artemisinin (also known as *qinghaosu*) and dihydroartemisinin. Her initial work was done in a secret laboratory and was designed to aid North Vietnamese soldiers during the war with the United States. She and her assistants reviewed more than two thousand recipes for traditional Chinese medicines before developing an effective strategy for extracting the drug from sweet wormwood (*Artemisia annua*), a common plant in China.[24]

Important advances in the methods needed to control malaria were made during the US occupation of Cuba, after the Spanish-American War, and during the construction of the Panama Canal. The beginning of the end of malaria in the United States came when malaria control was integrated into the Tennessee Valley Authority's mission. As a result, the disease is now uncommon in this country.

Malaria still poses a low-level threat. In its 2014 Malaria Surveillance Report, the CDC reported that there were 1,687 cases in the United States during 2012.[25] All but four of these were imported into the country. Although rare, transmission in the United States has been reported to occur due to infected mosquitoes; the potential is the highest in the South due to the abundance of potential vectors, so-called airport malaria caused by infected mosquitoes that hitchhike aboard aircraft, congenital malaria due to transplacental spread, and transfusion transmission.

All malaria parasites belong to a single genus, *Plasmodium*. According to the CDC malaria website, there are approximately one hundred different species in this genus; however, only five routinely cause human disease.[26] The most common species affecting humans are *P. falciparum*, found worldwide in subtropical and tropical areas, and *P. vivax*, found mainly in some parts of Africa, Asia, and Latin America. *P. ovale* is found mainly in Africa and islands in the western part of the Pacific. These are known as *tertian* forms of malaria, because fevers recur every forty-eight hours. *P. malariae* is found worldwide and is the only one of the group that has a three-day cycle, with fevers occurring every seventy-two hours (*quartan* malaria). *P. knowlesi* occurs in macaque monkeys, but it may infect humans and progress rapidly because of its twenty-four-hour replication cycle.

Malaria parasites exist in three stages: a mosquito stage, a human liver stage, and a human blood stage. Disease transmission occurs when an infected female mosquito bites a human, inoculating him or her with the parasites. The parasites travel to the liver and enter a liver cell, where they replicate and mature. Eventually, the infected liver cell ruptures, releasing the mature form of the parasite into the blood. The parasites then undergo asexual multiplication in red blood cells. The diagnosis of malaria can be made at this stage by examining appropriately stained red blood cells with a microscope. In time, the infected red cells rupture, releasing the parasites into the blood. Most clinical manifestations of the disease occur during this stage. The release of hemoglobin into the blood-stream can be massive and may cause severe anemia and kidney damage or failure. When hemoglobin is released into the blood stream, it colors the urine, a condition known colloquially as *blackwater fever*. (Note: Other disease states that cause red blood cells to rupture also can cause the urine to become dark in color.) When an uninfected mosquito bites an infected human, the mosquito ingests parasites and becomes infected. Another stage of the life cycle occurs in the mosquito, where parasites eventually mature into the form that can infect a human host during a blood meal.

This is a simplified description of the very complex malarial life cycle. Each stage in the cycle consists of several steps, and each of those steps is susceptible to modification. For those who are gluttons for punishment, more detailed descriptions, complete with the technical names for the parasite at various steps in each stage, can be found on the CDC malaria website mentioned earlier.

The mosquito stage is particularly susceptible to climate change and the dilution effect, described earlier. As the number of mosquito species decreases, possibly as the result of a changing climate, the probability of transmission rises.[27]

To better understand malaria and the factors affecting its transmission, it is necessary to be familiar with mosquitoes and some aspects of their life cycle. According to Wikipedia, there are over 3,500 different species of mosquitoes. Not all mosquitoes consume blood, and of those that do, pressure gradients between the host and the insect determine in part whether the mosquito has the capability to transmit disease. Only female mosquitoes consume blood. Mosquitoes have four stages of development: eggs, larvae, pupae, and adults. Most mosquitoes lay their eggs in water. Some are quite fussy and have a limited range of choices, and others are generalists and will lay their eggs almost anywhere. The elimination of breeding grounds has been of singular importance in malaria eradication. Therefore, an understanding of any preference or requirement is likely to be a critical element in mosquito control.

*Anopholes* mosquitoes, the vectors for malaria, breed in water and usually bite during the night. The introductory chapter referred to the life cycle of moths that live in the fur of sloths and the potential for many factors to disrupt this complex cycle: the same potential exists for

*Anopholes* mosquitoes. Water preferences vary, including any shallow collection, such as rice paddies, puddles, or even the hoofprints of animals. Thus, rainfall patterns, temperature, and humidity play major roles, which are all subject to the effects of climate change.[28] For most species of *Anopholes* mosquitoes, warm tropical temperatures and high humidity favor development and maturation of the parasites in the female insects. The risk of transmission is greatest at the end of rainy seasons, when mosquito eggs hatch and the larvae that live and develop in water mature into adults. The highest risk areas are those where weather or climate conditions change to favor mosquito reproduction and where the human population has a low immunity to the parasite.

Improvements in the control and reduction in the number of cases of malaria is one of the WHO Millennium Development Goals. Curiously, the 2013 progress report, while detailing many aspects of malaria control and threats to further progress, fails to address climate change as a threat.[29] However, the IPCC Working Group II report points out a number of climate-related events and circumstances that are likely to lead to increases in the risk of the development of malaria.[30] These challenges include large-scale disruption of populations due to the consequences of floods, rising sea level, and changes in precipitation; food insecurity, leading to undernutrition; and violence and the subsequent disruptions of social systems. Violence has been a severe problem in many parts of Africa, and undernutrition is a particular problem among children. Pregnancy is also cited as a risk factor for malaria. This increase in risk is thought to be due to changes in the immune system, formation of the placenta, and anemia. Disruption of the public health infrastructure due to the global financial crisis of the first part of this decade appears to have contributed to the decline in progress toward the eradication of malaria during the past several years, as fund-raising goals went unmet.[31]

In its analysis of vector-borne illnesses, the IPCC reports that there is *high confidence* that there is a positive association between temperature and humidity and malaria at a local level (where *high confidence* is a function of agreement among reports, their type, amount, quality, and consistency of the evidence).[32] In looking to the future, it is likely that malaria will spread into regions where it is not already present, while in other regions it may diminish. In some regions, malaria is already so prevalent that there is little or no room for an increase. Thus, it is possible that the global disease burden for malaria could change little but that there could be major changes in the distribution of risk.

## Water-Borne Illnesses

Many of the diseases in this category are transmitted by exposure to or the ingestion of contaminated drinking water or water used for bathing, washing, or swimming. Infections may also occur after exposure of a cut or other open wound or contact with eyes or ears. The oral–fecal route is particularly important in the transmission of these diseases, which include cholera. Other diseases are due to infections with the *Salmonella* and *Campylobacter* species of bacteria. Outbreaks of the latter two are frequently associated with warming weather.

### Cholera

Cholera is caused by the bacterium *Vibrio cholerae*. The disease manifestations are caused by toxins produced by the bacteria and not the bacteria themselves. There are over two hundred identifiable strains of the bacterium, but only two produce the toxin responsible for manifestations of the disease. Vibrios typically exist among phytoplankton. Small crustaceans known as *copopods* feed on plankton, and even though they may be only a few millimeters in size, a single copopod may contain a large enough number of bacteria to cause disease if ingested. Usually, however, the cholera bacteria are in an inactive state in the plankton. The threat of a disease eruption rises when algae bloom (proliferate). A single case of the disease can then give rise to many others, triggering an epidemic.

Because of links between temperature, precipitation, and time of year, several groups have been successful in defining relationships between the incidence of cholera and climatological variables. In Southeast Africa, the incidence of the disease increased by an annual factor of 1.08 between the years 1971 and 2006, an increase attributed primarily to global warming.[33] In a more complex study of cholera in Bangladesh, investigators found relationships between temperature and the number of hours of sunlight per day.[34] Increases in both were associated with increases in the number of new cases per month. However, since the cloud cover increased during the warmer summer, the synergistic effect was blunted. In another study of cholera in East Africa, the investigators found a significant interaction between temperature and rainfall amounts.[35] Each of these studies shows a significant relationship between climate and cholera, and, as a group, they help pave the way for predicting outbreaks. Armed with these data, public health officials may be able to better prepare for the future.

After a steady rise in the number of cases of cholera reported to the WHO in the beginning of this century, the number fell substantially in 2012.[36] During the intervening time, there was a shift in the burden of the disease from Asia to Africa. In 2012, two-thirds of all cholera deaths occurred on this continent. The Democratic Republic of Congo was hit the hardest that year with 33,611 cases and 819 deaths, a case fatality rate of 2.4 percent. The death rate in all of Africa was 1.7 percent, compared to a rate of 0.4 percent in Asia and 0 percent in Europe. Many of the deaths in the second decade of this century were the result of an epidemic in Haiti. This epidemic serves as an excellent example that illustrates the increased risk of cholera after natural disasters such as tropical storms and flooding, which are likely to occur more often due to climate change.

Cholera is characterized by nausea, vomiting, and severe diarrhea that may take on a milky appearance, so-called rice water diarrhea. In severe cases, victims may produce as much as five gallons of diarrheal fluid in a day, giving rise to severe dehydration, circulatory collapse, and death. The disease often progresses very rapidly, with only hours separating the onset of symptoms and death. Untreated severe cholera typically has a 50 percent mortality rate. However, with the widespread introduction of oral-rehydration therapy, the death rate can be below 1 percent. So-called cholera cots help prevent the spread of the disease in a healthcare facility by channeling feces into a receptacle. Measuring fecal volume and vomitus guides rehydration efforts and prevents volume depletion and shock. Antibiotics usually are not needed. This disease is most likely to occur in areas with poor sanitation and unsafe drinking water.

The Haitian epidemic began suddenly and explosively. The nation had been free of cholera for almost one hundred years.[37] That changed on October 18, 2010, when nine patients with diarrhea were hospitalized at the Mirebalis Hospital. Two days later, another nine were hospitalized. Three days later, thirty-five required hospitalization. A similar series of events unfolded at the Albert Schweitzer Hospital, where twenty-four patients were hospitalized on October 20. A total of five had been hospitalized during the preceding three days. Things were even worse at St. Nicholas Hospital, where 404 patients were hospitalized on October 20. There were forty-four deaths that day. On the day before, there were no hospitalizations for diarrhea. The Haitian National Public Health Laboratory rapidly determined that these patients were suffering from cholera. Within days, virtually the entire nation was at risk.

Although cholera persists in Haiti, the number of new cases has declined substantially. In a report issued at the end of June 2014, the Pan American Health Organization, part of the World Health Organization said there had been 703,510 cases of cholera in Haiti, including 393,912 hospitalizations and 8,562 deaths that were attributed to the epidemic.[38]

The conditions that led to the epidemic followed the devastating magnitude 7.0 earthquake that struck the nation on January 12, 2010. In spite of international aid from many nations, the infrastructure of this already poor nation was damaged severely, contributing to the conditions that led to the epidemic. To make matters worse, Hurricane Tomas struck the country on November 5. The ensuing flooding and damage to an already weakened country exacerbated the epidemic. In a final coup de grace, many of the climatic conditions in Haiti—including warm moist weather with critical mixing of seawater and freshwater at the mouths of rivers—favored spread of the disease.

Although there was a considerable amount of controversy concerning the origin of the outbreak, it seems virtually certain that it arose from a United Nations Stabilization Mission in Haiti (MINUSTAH) camp along the Artibonite River, where sanitation conditions were wretched and insufficient to protect river water. Molecular biological techniques were used to evaluate the bacterial isolates. These studies showed that the serotype of the bacteria was typical of the South Asian strains, proving that the bacteria did not originate in Haiti. The MINUSTAH camp in question was staffed by Nepalese; blaming them touched off a political storm. This appears to be an example of a disease being transported from one part of the world to another due to globalization.

Many of the recommendations of the Independent UN Committee that investigated the Haitian epidemic apply to preparing for a world affected by global warming and the expected increase in the severity of extreme weather events, such as tropical storms and coastal flooding. First, improving the infrastructure associated with delivering reliable, safe water supplies is essential. There also must be appropriate mechanisms in place for the disposal of human waste to prevent contamination of water used for drinking, bathing, and washing. Emergency responders must be prepared and screened appropriately to prevent the introduction of disease, as is suspected to have occurred in Haiti, and to prevent the outbreak of disease among aid workers.

### Hantavirus Diseases

Hantaviruses are a relatively new discovery. These viruses are carried by but do not cause disease in rodents. According to the hantavirus site maintained by the CDC, the rodents most likely to be carriers in the United States include deer mice, cotton rats, rice rats, and white-footed mice. Hantavirus disease gained attention during the Korean War, when around 3,200 UN soldiers developed a hemorrhagic fever that led to the isolation of the virus.[39]

There are two conditions linked to infection by the virus: Hemorrhagic Fever with Renal Syndrome (HFRS) and Hantavirus Pulmonary Syndrome (HPS). HFRS typically begins with relatively nondescript symptoms, including fever with chills, headache, back and abdominal pain, and nausea. The full-blown illness consists of five phases: the febrile phase; a hypotensive phase characterized by low platelet counts, low blood pressure, and low amounts of oxygen in the blood; an oliguric phase, with reduced urine output due to renal failure; a diuretic phase in which urine output may reach a gallon or more daily; and a convalescent phase. Death or permanent kidney damage may occur.

HPS, the form occurring most commonly in the United States, begins like HFRS. Those with the most serious manifestations rapidly develop shortness of breath, with chest x-ray evidence of adult respiratory distress syndrome. In spite of supportive therapy with oxygen and mechanical ventilation, the mortality rate is around 40 percent. HPS was identified near the Four Corners area of the United States when a man died after developing respiratory failure while on his way to his fiancée's funeral. She had died of a similar illness. Subsequent evaluations determined that he had a hantavirus infection.

These illnesses are much more common than many people appreciate. The most recent IPCC report estimates that about two hundred thousand patients are hospitalized each year due to hantavirus infections. Between 1951 and 1986, 14,309 individuals were hospitalized in Korea with the disease, although it is much less common in the United States: According to the CDC, there were 606 confirmed cases between 1993 and the end of 2013.

The 1993 outbreak and subsequent cases identified more recently among visitors to Yosemite National Park led to an intensification of efforts to study the disease. Once again, disease ecologists have made important contributions to such studies.[40] Their investigations began with the observation that the rodents carrying the disease were much more common after periods of above-normal precipitation in desert areas. They then combined data from vegetation maps derived from satellite data, hydrological data from weather stations, and geological information (such as elevation and the slope of the terrain) to make highly accurate predictions of the rate of infection in deer mice. It also seems likely that, once again, patterns of precipitation in the southwest desert are dependent on El Niño and the Southern Oscillation and are related to increases in the risk of hantavirus diseases.

## Leishmaniasis

Leishmaniasis is a disease caused by an infection by any one of approximately twenty species of protozoa from the genus *Leishmania*. WHO estimates that there are around 1.3 million new cases and between 20,000 and 30,000 deaths each year. There are three main forms of leishmaniasis. The most serious, visceral leishmaniasis, kala-azar, is characterized by fever, enlargement of the liver and spleen, and anemia. It is fatal if untreated. The vast majority of new cases (about 90 percent) occur in Bangladesh, Brazil, Ethiopia, India, Sudan, and South Sudan. Cutaneous leishmaniasis is the most common form and causes ulcers on the skin that leave scars and may cause serious disability. About two-thirds of the cases occur in Afghanistan, Algeria, Brazil, Columbia, Iran, and Syria. Mucocutaneous leishmaniasis causes destruction of mucous membranes in the nose, mouth, and throat. Bolivia, Brazil, and Peru harbor most of these cases. It is rare in the United States, with most cases contracted during visits to areas where it is endemic.

The disease is transmitted by the bite of infected female sand flies (more specifically, phlebotomine flies). A number of animals act as reservoirs for the parasite. The range of the reservoirs and flies is affected by temperature, rainfall, and humidity. Thus, the disease is likely to be affected by climate change. This issue has been addressed. The risk of it spreading to North America, where increases in the range of the disease are predicted as climate change becomes more severe, was discussed in a 2010 report.[41] Recent research has shown that sand flies benefit from the leishmanial infection by gaining resistance to pathogens that affect them.[42]

The disease can be diagnosed by observing the parasites with a microscope in samples of white cell–enriched blood or bone marrow. It can be treated with antibiotics, particularly lysosomal amphotericin B. Preventive measures include sleeping under insecticide impregnated nets.

## The Trajectory toward the Future

Large numbers of people worldwide have diseases, as described in the preceding sections, that are likely to be affected by climate change. Many, but not all, are tropical in their distribution, and their range may expand in a warmer world. Many of these diseases are *zoonotic*; that is, they are transmitted from other animals to humans. Thus, understanding the future impacts of these diseases depends on an understanding of the effects of climate on the animal host and the organism responsible for causing the disease.

Disease ecology is likely to become increasingly important, particularly for diseases with limited or poor treatments. The study of the interactions between climate variables, such as temperature and precipitation; disease vectors, such as mosquitoes; pathogens, such as malaria parasites; and how changes in climate affect disease patterns will become increasingly important. These studies have been aided enormously by the development of remote sensing equipment, often carried by satellites. These data, combined with mathematical models that link climate variables to disease patterns, have already led to improvements in public health. For example, armed with the results of these models it should be possible to mobilize the resources needed to combat predicted outbreaks or increases in the risk of contracting a specific disease. Seemingly remote changes in local conditions that are associated with El Niño and the Southern Oscillation can be used to predict changing patterns of disease.

Using this knowledge effectively may be more difficult than acquiring the knowledge in the first place. As discussed in the first chapter, lack of political will, reluctance to mobilize resources to help distant populations, adverse local conditions due to political considerations, and violence are almost certain to create barriers to developing measures that are needed to prevent and adapt to climate change.

# 5

# Climate Change, Agriculture, and Famine

Climate change poses unprecedented challenges to US agriculture because of the sensitivity of agricultural productivity and costs to changing climate conditions.

—*USDA Technical Bulletin 1935*, 2012

Other chapters described some of the effects of climate change that impact humans directly: the effects of heat, disease, and so on. Although there are genetic differences among us that make us more or less susceptible to various diseases, we are all one species. We have similar, but not identical, sensitivities to the stresses of all types that are expected as the result of climate change. This chapter turns to agriculture, the cultivation of crops, animals, and other forms of life that provide the food, fuel, and other needs of civilization. In agriculture, it is necessary to consider multiple species, not just humans. Not surprisingly, the differences among species are immense—and so are the responses to a changing climate.

The ability to grow adequate amounts of food is a prerequisite to good health. For each of the species that make up the total output of crops and animals that constitute agriculture, there are defined optimal and limiting temperatures for growth, reproduction, and development. Different species have vastly different susceptibilities to numerous diseases, insects, and weeds; different needs for water; and different responses to the increase in atmospheric concentration of $CO_2$. The concentration of this long-lived greenhouse gas is rising and is certain to continue to rise. We will likely require different mitigation and adaptation strategies for each component of agriculture and for agriculture as a whole.

The task of providing adequate nutrition for the growing number of the earth's inhabitants would be daunting even without climate change.

We already fail to feed everyone adequately. The 2014 publication titled *Hunger Report: Ending Hunger in America*, published by the Bread for the World Institute, reports that between 2010 and 2012, 14.7 percent of Americans had an insecure food supply, and that in 2012, over forty-six million relied on the Supplemental Nutrition Assistance Program (SNAP) to feed themselves and their families.[1] Many of those who are hungry are children.

Although there are fewer people who suffer from malnutrition than in the past, the problem is still one of "staggering size," as reported in the 2015 Global Nutrition Report published by the International Food Policy Institute.[2] The institute found that malnutrition affects all countries and one in three people on the planet. Malnutrition, particularly when it takes the form of undernutrition, results in stunting. Stunting is present when a child is two standard deviations below World Health Organization median values for either height or weight for age. This translates into a bit more than the lowest two percent of the weight that separates the top half from the bottom half of children at any given age. Undernutrition and stunting have profound effects on the developing brain. The result is a lifelong impairment of brain function, including intelligence. This decreases the probability that individuals and societies will thrive. The World Bank reports that in some countries in the years between 2009 and 2013, 45 percent of children under five years of age did not receive adequate nutrition. The effects of the expected changes in our climate will increase the difficulties associated with feeding everyone and could make problems associated with undernutrition much worse. The challenge is enormous.

Climate change already affects agriculture in this country. We are one of the leading producers of agricultural commodities. According to the Food and Agriculture Organization of the United Nations, the United States produces about 35 percent of the world's corn, 32 percent of the soybeans, and smaller but significant amounts of other leading crops such as rice and wheat, about 1 percent and 8 percent, respectively. What happens here has already affected, and will continue to affect, others around the world. For additional details on crop production, see table 5.1.

Table 5.1
Agricultural commodity production, 2013, in millions of metric tons

| Crop | US yield | World yield |
| --- | --- | --- |
| Maize | 353.7 | 1,016.7 |
| Rice | 8.6 | 745.7 |
| Soybeans | 89.5 | 276.4 |
| Wheat | 58.0 | 713.2 |

*Source:* Food and Agriculture Organization of the United Nations, http://www.fao.org/docrep/018/i3107e/i3107e00.htm.

## Effect of Atmospheric $CO_2$ on Plant Growth

"Carbon dioxide is plant food, bring it on!" So goes the cry of climate change deniers. There is some truth to what they say. However, like virtually every other detail about climate change, it is not nearly that simple.

Six hundred million years ago, when the ancestors of the plants we recognize today began to develop, the earth's atmosphere contained huge amounts of $CO_2$.[3] Eventually, as plants proliferated, carbon dioxide levels fell as a result of the combined effects of a reduction in natural emissions of the gas and carbon trapping by plants. When these plants died, many were buried and eventually formed the fossil fuels we extract and burn today. Although there were fluctuations, atmospheric levels of $CO_2$ were still high until just over twenty-four million years ago, when they fell to around 300 ppm. This concentration remained relatively constant until the onset of the industrial revolution (see chapter 2).[4] From an evolutionary perspective, plants evolved during periods of high concentrations of $CO_2$. Compared to the conditions that were present when they first appeared, today many plant species are relatively starved for $CO_2$. Many plants would like more of the gas and would grow more rapidly with a higher $CO_2$ concentration.

Virtually all plants of interest fall into one of two groups: those that respond to increases in the concentration of $CO_2$ and those that do not. Which group a plant falls into depends on how its photosynthetic pathways function. About 95 percent of plants produce a three-carbon compound during the initial step of this process that converts $CO_2$ into sugars.

These are referred to as $C_3$ plants. Soybeans, rice, and alfalfa are good examples of $C_3$ plants. Plants in this group are responsive to changes in the concentration of $CO_2$. Typically, their growth rates increase as $CO_2$ concentrations rise, at least in short-term studies. The less common $C_4$ plants produce a four-carbon compound during the initial step of photosynthesis. Corn and sorghum are examples. $C_4$ plants do not respond to increases in $CO_2$ concentrations because the enzymes in their metabolic pathways are saturated with the gas, and therefore increases have no effect on growth. Think of people trying to get into a bank via a revolving door. When there are only a few people (molecules of $CO_2$) trying to get in, an increase in the number of those trying to enter results in an increase in the number getting into the bank. This is the $C_3$ situation: adding more $CO_2$ (people) results in an increase in photosynthesis. However, when there are crowds of people outside the bank, the revolving door can't let more people in; the door is saturated. This is the $C_4$ condition.

Most short-term research shows that plants grow more rapidly and accumulate mass as the ambient $CO_2$ concentration increases.[5] To illustrate this, consider a study of grasslands in Texas that contained a mixture of $C_3$ and $C_4$ species from which cattle had been excluded.[6] Four years later, this plot was exposed to a gradient in the concentration of $CO_2$ that ranged from 200 ppm to 560 ppm. From chapter 2, you may remember that the ambient $CO_2$ concentration is now about 400 ppm. Researchers found that the aboveground biomass increased between 121 and 161 grams per square meter for a 100 ppm rise in the $CO_2$ concentration. Belowground biomass was not measured. In response to the added $CO_2$, the composition of the grassland shifted from $C_4$ grasses to $C_3$ flowering plants that are not grasses (technically, the shift was to *forbs*, flowering plants with leaves and stems). Neither the grasses nor the flowering plants were more desirable; they were just different.

This study of Texas grasslands is representative of what might be expected to happen in a mixed plot of $C_3$ weeds and $C_4$ crops as $CO_2$ concentrations increase. Theoretically, when a $C_3$ weed invades a $C_4$ crop, the weed is more likely to win when the $CO_2$ concentration rises. Table 5.2 shows some of the winners and losers in some pairings of crops and weeds of interest. Under laboratory conditions, it has been shown that $C_3$ weeds often muscle out $C_4$ crops, resulting in a decreased yield of the crop.[7]

**Table 5.2**
Crops vs. weeds

| Crop | Weed | High CO$_2$ favors |
|------|------|--------------------|
| **C$_4$ Crop–C$_4$ Weed** | | |
| Sorghum | Redroot amaranth | Weed |
| **C$_4$ Crop–C$_3$ Weed** | | |
| Sorghum | Rough or common cocklebur | Weed |
| Sorghum | Velvetleaf | |
| **C$_3$ Crop–C$_4$ Weed** | | |
| Fescue | Johnson grass | Crop |
| Soybean | Johnson grass | Crop |
| Rice | *Echinochloa glabrescens* | Crop |
| Soybean | Redroot amaranth | Crop |
| **C$_3$ Crop–C$_3$ Weed** | | |
| Soybean | Creeping thistle | Weed |
| Soybean | Goosefoot | Weed |
| Alfalfa | Dandelion | Weed |
| Pasture | *Plantago* (a genus with 200 species—e.g., plantain) | Weed |
| Pasture | English or narrowleaf plantain | |

*Source:* Adapted from C. L. Walthall, P. Hatfield, L. Backlund, et al., *Climate Change and Agriculture in the United States: Effects and Adaptation; USDA Technical Bulletin 1935* (Washington, DC: United States Department of Agriculture, 2012), table 4.1.

An important and convincing challenge to the "CO$_2$ is fertilizer" position, touted by climate change deniers, arises from a recent study of trees in the Amazon River basin.[8] The study was designed to answer the question "Does a rising atmospheric CO$_2$ concentration cause trees to trap more of this greenhouse gas?" In a massive undertaking, teams of scientists made direct measurements of all trees with a diameter of 100 mm or more in 321 different plots of land scattered throughout the Amazon basin. Candidate plots were excluded if there was evidence for recent human activity within their boundaries. In addition, the plots were selected to provide lots of diversity to make them as representative as possible of the many different ecosystems in the entire basin. Periodic measurements began in 1983, when twenty-five plots were marked out, and continued into the middle of 2011, when the full number of plots was under surveillance. Tree size data from serial measurements in the plots were used to compute the amount of carbon that had been trapped, and trapped carbon data for all trees were summed for all of the plots.

During the study, the atmospheric $CO_2$ level rose steadily. The results of the investigation were striking. Individual trees did indeed grow more rapidly in response to the rising $CO_2$ and, accordingly, trapped more $CO_2$. However, this effect was overwhelmed by the fact that trees were not living as long as they had in the past. This resulted in a net decrease in the rate at which $CO_2$ was trapped. The effect was not trivial. The authors of the paper wrote, "The above-ground biomass declined by one-third during the past decade compared to the 1990s."[9] When these data are combined with the fact that massive amounts of tropical forests are cleared each year, there is real cause for concern about the ability of the natural system to mitigate climate change.

In her book *The Sixth Extinction: An Unnatural History*, Elizabeth Kolbert paints a vivid picture of the diversity to be found in the Amazon basin.[10] As she was guided through the plots, similar to those described previously, she was told to look carefully at the leaves on the plants and trees. As Kolbert climbed a hill, her guide told her that after walking a few hundred feet she would not see these same leaves again. The diversity was so enormous and the plant life was so highly adapted to a narrow range of temperatures and humidity that small changes in elevation moved an observer into an entirely new ecosystem. Although the authors of the study documenting the decline of the Amazonian carbon sink did not reference specific species in their report, it seems quite likely that as $CO_2$ levels rise, the diversity in the ecosystem will fall, just as was seen in the Texas prairie study. This will make the Amazon even more susceptible to the effects of climate change. A loss of diversity may be as disastrous as the impact on the carbon sink.

In addition to the atmospheric $CO_2$ concentration, plant growth depends on many other factors. The amount of nitrogen in the soil is critical. This is why farmers add nitrogen to their fields and homeowners apply nitrogen-rich fertilizers to their lawns. As you might expect, the balance between nitrogen and $CO_2$ is important. In a report examining the interaction between these two chemicals, the authors noted that most of the studies that examined the effect of $CO_2$ were of a relatively short duration. Therefore, they embarked on a six-year investigation that examined the $CO_2$–nitrogen relationship and found, as the title of their paper states explicitly, that nitrogen limitation constrains sustainability of ecosystem response to $CO_2$.[11] The nutrient content of the soil for plants grown under conditions of an elevated concentration of $CO_2$ is arguably more important than whether the plant grows faster. In other words, the $CO_2$ fertilization effect is subject to considerable modification by the nitrogen concentration in the soil. This is likely to be most important in parts of the world where (1) the soil condition is poor because the same crop is planted year after year, (2) the soil is damaged, and (3) farmers can't afford chemical fertilizers. This is the case in much of sub-Saharan Africa. (Chapter 10 describes efforts to adapt to and overcome the poor soil condition in parts of Africa.)

In addition to the complex relationship between $CO_2$ levels and plant growth, the nutritional content of some crops is also affected by $CO_2$. Inadequate nutrition is an important determinant of the global burden of disease, as discussed in chapter 1 and shown explicitly in table 1.2. Therefore, more in-depth studies of the numerous effects of climate change on crops are warranted. Limited amounts of iron and zinc are particularly problematic in many parts of the world. In a recent study that combined new data with that which had already been published, the study authors noted that around two billion people are deficient in zinc and iron.[12] They reported that increasing the concentration of $CO_2$ in the ambient air led to a reduction in these two critical elements in $C_3$ grains and legumes. These two classes of food supply most of the micronutrients in regions where dietary deficiencies are rampant. The study authors also found reductions of protein in many $C_3$ crops other than legumes. Finally, they found variable effects of $CO_2$ on the concentration of a molecule (phytate) that inhibits the uptake of zinc in the human gastrointestinal system. Phytate levels were reduced in wheat, a $C_3$ grass. This might mitigate the effects of reduced zinc. Plant phytate concentrations are used to model zinc metabolism, so a thorough understanding of the behavior of this molecule is important. Some of the study results are shown in figure 5.1.

To summarize, although it is clear that some plants (notably the $C_3$ species) grow more rapidly in response to an increase in the $CO_2$ concentration, this increase comes with a price. Protein yields in these plants may be low, other nutrients may be reduced, and the $CO_2$ fertilizer effect may not be permanent. Under conditions of an elevated $CO_2$ concentration,

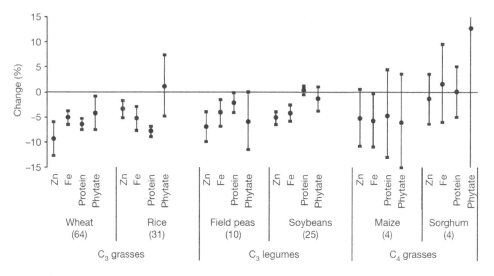

**Figure 5.1**

Percentage change in nutrients at elevated $CO_2$ concentrations relative to the ambient $CO_2$ concentration. Numbers in parentheses refer to the number of comparisons in which replicates of a particular cultivar grown under one set of growing conditions in one year at elevated $[CO_2]$ have been pooled and for which mean nutrient values for these replicates are compared with mean values for identical cultivars under identical growing conditions, except grown at ambient $[CO_2]$. In most instances, data from four replicates were pooled for each value. Error bars represent 95 percent confidence intervals of the estimates, and $[CO_2]$ represents the atmospheric concentration of $CO_2$. Reproduced with permission from S. S. Myers, A. Zanobetti, I. Kloog, et al., "Increasing $CO_2$ Threatens Human Nutrition," *Nature* 510, no. 7503 (2014): 139–142.

plant diversity may fall and weeds and invasive species may flourish. In other words, when something looks like it is too good to be true, it probably is.

**Ragweed**

Ragweed (*Ambrosia artemisiifolia*) is common throughout North America and can be found in Europe. Its Latin name, suggesting a relationship to the nectar of the gods and to asters, gives this plant an undeserved positive spin. It produces pollen that is highly allergenic and acts as a potent trigger for attacks of asthma. It is also responsible for much of the suffering among those afflicted by hay fever. The lengthening of the grow-ing season and increases in the atmospheric $CO_2$ concentration that char-acterize climate change favor the growth of this pest.

Plant scientists interested in ragweed have grown it under strict labo-ratory conditions.[13] In one of the most comprehensive studies, a group of scientists tested the hypothesis that a longer growing period and ele-vated $CO_2$ levels would lead to the production of more of the offensive pollen. They released dormant ragweed seeds during three successive fifteen-day intervals. At each of the three releases, the $CO_2$ concentration was identical to the level in the atmosphere or increased to 700 ppm. The heights and weights of the plants were measured periodically along with the number, length, and weight of the clusters of flowers and the time at which the flowers opened. The scientists placed bags over the flowers to quantify pollen production. The seeds planted in the first group outgrew those planted later by almost every measure. They were bigger and produced more clusters of flowers that were heavier and yielded almost 55 percent more pollen than seeds planted in the third cohort. These results were further augmented when plants were grown in the $CO_2$-enriched atmosphere. This interaction between the longer growing season and enhanced pollen production at high $CO_2$ concentra-tions predicts more suffering for asthmatics and hay fever patients as the climate warms.

### Effects of Temperature on Plant Growth

The relationships between the ambient temperature and plant growth are complex. Each species has a temperature below which growth does not take place and a temperature above which growth fails. In between, there is an optimum temperature. In addition to growth, temperature affects pollination and other aspects of plant reproduction. Thus disruptions, particularly warming, may have substantial impacts on plants—including corn and soybeans, the most prevalent and valuable crops grown in the United States.

Corn is an important source of food for Americans, particularly when one considers the prevalence of corn-derived products such as high fructose corn syrup in contemporary diets and so-called nutraceuticals (nutrients and dietary supplements, as well as a variety of foods that may or may not have direct links to corn). Corn is also the primary source of carbohydrates that are used in fermentation reactions to produce the ethanol that makes up 10 percent of the gasoline mixture at the pump. In addition, farmers who raise hogs, poultry, and cattle, both for meat and dairy production, rely heavily on corn and soybeans for feed.

Several temperatures are important when discussing the relationship between temperature and yield. The first of these is the *optimum temperature for grain yield*, which the US Department of Agriculture lists for a number of crops. For corn, it is between 18°C and 22°C; for rice, it is somewhat higher, between 23°C and 26°C; and for wheat, it is lower at 15°C.[14]

An extensive analysis of the impact of temperature on corn and soybean yields was published in a 2009 report.[15] For both crops, the yield increases slightly as degree-days increase up to a critical level (where a *degree-day* is a function of the average temperature on a given day). Beyond a *critical temperature*, yields fall—and fall rapidly. For corn, the critical temperature is 29°C, and it is 30°C for soybeans. There is likely to be a major impact on agricultural production as daily temperatures increase above critical levels in parts of the country where these are the dominant crops. This is particularly true for the Midwestern part of the United States and for Africa.

As temperatures continue to rise, farmers may encounter the *failure temperature*. This is exactly what it sounds like—the temperature that results in failure of the crop, not just a reduction in yield. Frequently, this failure occurs during pollination, the most temperature-sensitive portion of crop production. The US Department of Agriculture also lists failure temperatures for various crops. For beans, it is 32°C; for wheat, it is 34°C; for rice, sorghum, and corn, it is 35°C; and for soybeans, the failure temperature is 39°C.

The 2009 report on temperatures and crop production made yield projections based on climate models commonly in use at the time.[16] The most optimistic of these, the B1 Hadley III warming scenario, predicts a temperature increase of about 1°C above current temperatures at the end of the century. This is expected to be associated with a yield reduction of approximately 45 percent for corn and 35 percent for soybeans. The least optimistic scenario, the A1FI Hadley warming scenario (where *FI* indicates *fossil fuel intense*—i.e., burning lots of fossil fuels) predicts an increase of about 3.5°C above current temperatures by the end of the century. At this level of warming, scientists predict that there will be an 80 percent reduction of corn yields and slightly lower reductions for soybean yields.

These data are critical predictors of the future, but evidence of what has happened already is perhaps even more important. Databases that are available to the public are extremely useful for this purpose. One such study was published in 2011 and has been widely cited.[17] For all nations, the authors retrieved information on crop locations, monthly temperatures, and precipitation for four key crops: maize or corn, wheat, rice, and soybeans. These crops were chosen in part because they account for three-fourths of global consumption of calories. Temperatures were relatively stable between 1960 and 1980. However, for the next two decades temperatures warmed and varied substantially, as discussed in chapter 2. During those decades, 75 percent of all countries had a one-standard-deviation increase in temperature trends in growing regions for wheat, 65 percent had an increase of a similar magnitude in regions growing maize and rice, and 53 percent had an increase in regions growing soybeans.

The authors compared actual yields to those that were predicted by modeling in the absence of a temperature change. The observed global yields for maize and wheat production fell by 3.8 percent and 5.5 percent, respectively. Among variations in rice and soybean production, temperature changes had little effect on worldwide yields. However, as might be expected, there were larger effects observed in some countries. For example, maize production in Brazil and China fell by about 7.5 percent, wheat production in Russia was depressed by almost 15 percent, and soybean production was down by about 4 percent in Brazil and Paraguay. The authors pointed out that any expected increases in yields due to rising $CO_2$ and improvements in technology were obliterated by the effects of climate change.

## Other Factors Affecting Agriculture

Thus far, I have focused on the effects of an increasing atmospheric $CO_2$ concentration and temperature in this chapter. These variables have major effects on agriculture and occur with some degree of evenness and predictability worldwide. However, climate change will also have nonuniform effects that are certain to affect crop production. These include the amount and intensity of precipitation, the prevalence of severe storms, changes in the concentration of ozone, proliferation of insect pests and impacts on insect pollinators, the effects on plant pathogens (yes, plants get sick too!), soil degradation, the effects of wind, and others. Most of these effects are worthy of books themselves, and details are beyond the scope of this chapter. Interested readers should consult the US Department of Agriculture 2013 publication *Climate Change and Agriculture in the United States: Effects and Adaptation*, an excellent and objective source.[18] However, it is worthwhile to consider the impacts of drought on agriculture in this chapter—something that seems self-evident.

Droughts have been the nemesis of the farmer throughout recorded history. This continues to the present day. In his book *Collapse: How Societies Choose to Fail or Succeed*, Jared Diamond illustrates the importance of drought in the failure of the Indian tribes of the Southwest, Mayan civilizations, and others.[19]

Major droughts had important impacts on agricultural productivity and water use in 2012 and 2014. The 2012 drought was the worst to occur in the midsection of the United States since the 1930s. Real and anticipated crop loss in the United States led to rapid and substantial worldwide increases in the price of food. On August 30, 2012, the World Bank reported that world food prices had increased by 10 percent in July due to the Midwestern drought. An analysis published by Bloomberg in January 2013 said that the drought had severely depleted soil moisture levels, as indicated by the Palmer Index, a widely used measure of drought severity.[20] Experts quoted in the report predicted that it might take between eighteen and fifty-one months to make a complete recovery, even if normal amounts of rain fell. At the time of this writing, an ongoing drought has affected major portions of California's Central Valley and adversely impacted food production for much of the nation. In response, Governor Brown of California announced rules to curtail water use substantially in nonagricultural regions of the state. Water shortages throughout much of the Southwestern part of the country, including the Rio Grande River system, have forced curtailment of water-intensive activities.

Sub-Saharan Africa has also been hard hit by drought. This part of the world was already economically disadvantaged and, in some areas, ravaged by violence. South Sudan was particularly devastated. On April 3, 2014, a UN official told the *New York Times* that 3.7 million South Sudanese people, or one-third of the population, was on the verge of starvation.[21] He predicted that unless $230 million in food aid was forthcoming within a two-month window, deaths due to starvation could rival those from the 1980s drought, when hundreds of thousands died in Ethiopia.

**Famine**

Famine frequently follows in the footsteps of drought. In the summary findings of a 1996 policy research working paper, the author defined famine as "widespread, usually life-threatening hunger or starvation."[22] The author listed the following risk factors that increase the vulnerability of a region to famine: poverty, a weak social and physical infrastructure, a weak and unprepared government, and a relatively closed political regime.

In the Key Risks section of the Summary for Policymakers, the IPCC Fifth Assessment Report stated with high confidence that there is a high "risk of food insecurity and breakdown of food systems linked to warming, drought, flooding, and precipitation variability and extremes, particularly for poorer populations in urban and rural settings."[23] The current state of nutrition leaves much to be desired; figure 5.2 shows the percentage of undernourished people in various regions of the world. The 2014 UN Food and Agriculture Organization report estimates that between 2012 and 2014 there were 805 million people who were undernourished.[24] This number, while still too large, represents a reduction of more than one hundred million compared to the number affected in the 1990 to 1992 interval. The report goes on to say that although sixty-three nations reached the Millennium Development Goal for hunger abatement large portions of the world—including sub-Saharan Africa, the Caribbean, Southern Asia, and Oceania—have not. Oceania is the region with the smallest population, but the absolute number of hungry individuals has increased, and rising rates of obesity, considered by some to be a form of malnutrition, are problematic.

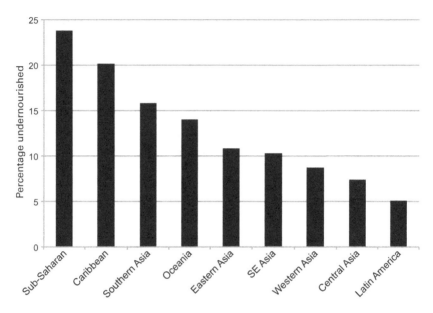

**Figure 5.2**

Undernourishment by region. The percentage of people in each region who are undernourished, adapted from FAO, IFAD, and WFP, *The State of Food Insecurity in the World 2014: Strengthening the Enabling Environment for Food Security and Nutrition* (Rome: FAO, 2014). Millennium Development Goals remain unmet in sub-Saharan Africa, the Caribbean, Southern Asia, and Oceania.

## Famine and Violence

On December 17, 2010, Tarek al-Tayeb Mohamed Bouazizi, a young Tunisian street vendor, set himself on fire to protest alleged harassment and confiscation of his goods by local authorities. This act, which garnered worldwide attention, came at a time of rising food prices and led to demonstrations and to what is now referred to as the Arab Spring, which reached its culmination with the downfall of the Tunisian and later the Egyptian governments.

The link between food and violence is not new. An analysis of food and violence in North Africa and the Middle East resulted in a model that may predict unrest from food.[25] Peaks in the UN Food and Agriculture Organization's Food Price Index that coincide with food shortages demonstrate a remarkable concordance with the timing of food riots that occurred in that region, which suggests that adequate amounts of food and national security and tranquility may be intimately linked. For additional details concerning food riots, refer to chapter 8.

## The Trajectory toward the Future

This chapter only begins to scratch the surface of the extraordinarily complex relationship between climate change, agriculture, and the ability of the world to feed itself. In his pessimistic 1798 publication *An Essay on the Principle of Population*, Thomas Malthus wrote that if disease and pestilence did not kill us off, "gigantic inevitable famine stalks in the rear, and with one mighty blow levels the population with the food of the world."[26] Thus far, with notable exceptions, we have avoided this Malthusian catastrophe. There have been exceptions, such as the Anasazi, ancestors of the Pueblos, who are often thought to have been driven from their land by periods of climate instability and drought. It remains to be seen whether our increasingly globalized society will be able to feed itself in the face of climate change.

# 6

# Sea Level Rise and Environmental Refugees

Our people will have to move as the tides have reached our homes and villages.

—Anote Tong, President of Kiribati, in talks with Fiji about moving his entire nation to a new site[1]

In March 2012, President Anote Tong of Kiribati began negotiations with the government of Fiji to purchase land on Vanau Levu. His island nation will need a new a home after it is inundated by the rising seas caused by climate change. The leaders of Fiji recognize that such a move will be difficult, because "they are going to leave behind their culture, their way of life, and lifestyle." Technically, this will be a planned migration, but make no mistake: the people of Kiribati will be refugees, among the many who will be displaced by a global increase in the sea level.

The vulnerability of Kiribati, Vanuatu, and other islands in the Gilbert chain, located north and east of Australia, was shown clearly by the arrival of Cyclone Pam in March 2015. This intense storm packed winds of 165 mph (270 kph) and devastated the islands, and the residents of these islands are not alone: this is but one example of the problems that face people who live close to the ocean.

A 2007 report on the risks of climate change to those living in low-elevation costal zones provided insight into the scope of the problem.[2] The report's authors found that 2 percent of the earth's surface is ten meters or less above sea level. However, because people tend to live in coastal areas, this area is home to 10 percent of the world's population and 13 percent of all urban dwellers. Although many of the countries in the survey are island nations like Kiribati, most countries with large populations have a high concentration of individuals living on river deltas that are at or near sea level. The IPCC Fourth Assessment Report singled out the Nile, Ganges-Brahmaputra, and Mekong deltas because of their *extreme* vulnerability to flooding. They classified the Mississippi Delta as a region of *high* vulnerability. These delta regions have become vulnerable because of the combined effects of (1) subsidence due to pumping out water and petroleum located beneath them, (2) reductions in delta maintenance due to dam blockage of sediments and other factors, and (3) increases in sea level due to climate change.

## Sea Level

Initially, measuring the level of the ocean seems like a simple task. One only needs to place a measuring device on a post in the ocean and record the change in water level at any given moment. For centuries, this is how it was done. The average result was recorded as the mean sea level. This is in accord with the IPCC glossary, which defines mean sea level as "the surface level of the ocean at a particular point measured over an extended period of time."[3] These measurements were made with tide gauges placed at various points around the earth until 1993, when these tide gauge measurements were augmented by devices carried on a series of satellites.

Satellite-derived measurements are based on the round-trip time of radar or laser energy emitted by the satellite as it is reflected back to the satellite and a knowledge of the satellite's position, and these measurements depend on GPS technology. The change from tide gauges to satellites results in a difference in the reference point for the measurements. Tide gauges use the land surface as the point of reference, whereas satellites use the immobile center of the earth as the point of reference.

Satellite measurements produce what is known as the *geocentric mean sea level*. Satellites are also used to measure the elevation of land, sea ice, the snow over sea ice, seawater, and clouds. Tide gauge measurements are affected by a rise or fall in the water, a rise or fall in the land, or both. Thus, tide gauge and satellite-based measurements are not always the same. For example, tide gauge measurements made in Stockholm, Sweden, show a fall in mean sea level. This is due to an uplift of the land that is occurring due to the absence of the pressure of ice on the earth's surface during the last ice age, some twenty thousand years ago. In other areas, such as Manila in the Philippines, the land is sinking because large amounts of water have been pumped out of the ground. In a literature review, a 60 cm increase in geocentric mean sea level would result in sea level increases measured with tide gauges of 70 cm in New York City, 88 cm at Hampton Roads, Virginia, and 107 cm at Galveston, Texas, where particularly large amounts of water and petroleum products have been removed.[4]

A closer look at the issue reveals some additional complexities that may seem trivial until you consider the vastness of the oceans. Sea level will change if the shape of the ocean floor changes, if the amount of water in the ocean changes, if the density of the water in the ocean changes, or due to some combination of these factors. These changes are referred to as *steric changes*, which may be due to changes in the temperature of the water (thermosteric) or due to changes in salinity (halosteric). Halosteric or thermosteric changes can be local or global or both. Melting ice sheets reduce the salinity and hence the density of water, as does heating. To this fact, add the effects of ocean currents, winds, periodic oscillations such as El Niño, deformations of the earth's crust due to tectonics, and other elements. In most cases, when discussing sea level changes, authors refer to *mean sea level* or *global mean sea level*.

### Changes in Sea Level

During the past 120 million years, there have been huge changes in the mean sea level. Data from the analysis of oxygen isotopes in fossils and other sources have shown that sea levels in the past were as much as 150 meters higher or lower than at the present time, depending on how much ice covered the earth.[5] These data are an indication of how much potential there is for future sea level changes.

More pertinent to modern life, measurements of sea level made since the late nineteenth century demonstrate an increase of almost nine inches, as shown in figure 6.1. An analysis of the tide gauge data shows that the rate of sea level rise was $1.7 \pm 0.3$ mm per year between January 1870 and December 2004. The scientists who recorded and analyzed these data restricted their inclusion of measurements to those made at geologically stable sites—that is, sites where neither the subsidence nor the elevation of the land was significant. In that interval, there was a significant acceleration of the rise in sea level of $0.013 \pm 0.006$ mm per year.[6] The validity

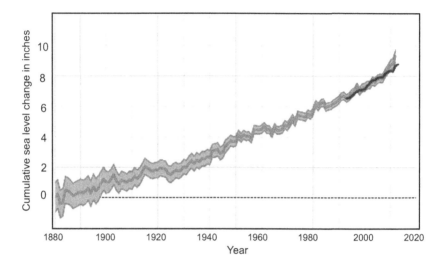

**Figure 6.1**

Global average absolute sea level changes, 1880–2013. Cumulative sea level changes based on tide gauge measurements, and likely range of values based on number of measurements and methodological precision (grey) and satellite altimetry (black). *Source:* EPA, "Climate Change Indicators in the United States," accessed October 24, 2014, www.epa.gov/climatechange/science/indicators/oceans/sea-level.html.

of this measured acceleration of the rate of sea level rise has been confirmed by satellite altimetry data collected between 1993 and 2003 that show a 3.1 mm per year rate of increase.[7]

The most recent IPCC report concludes, with high confidence, that 75 percent of the rise in sea level in the past decades is due to a combination of an increase in the volume of existing sea water due to heating (thermal expansion) and the addition of water to the oceans due to melting glaciers, as shown in figure 6.2. On average, glaciers were about twelve meters thinner in 2005 than in 1960. The melting of glaciers and thermal expansion of the oceans' waters have both been caused by human activities that have caused climate change. If the climate warms sufficiently, glaciers will disappear completely and will no longer contribute to increases in sea level. However, thermal expansion of the oceans will continue for very long periods, even if temperatures stabilize.[8] The inertia in this system is due to the fact that the land, sea-surface, and air

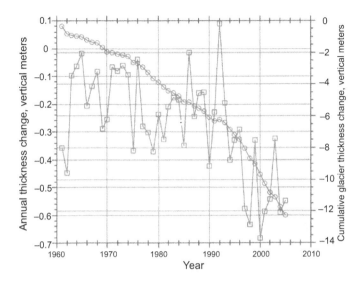

**Figure 6.2**

Average yearly and cumulative thickness of mountain glaciers. In most parts of the world, glaciers are shrinking in mass. Between 1961 and 2005, the thickness decreased by approximately twelve meters. *Source:* National Snow and Ice Data Center; image in public domain, modified from Wikimedia Commons (Wikipedia entry: Future Sea Level).

temperatures are not in a state of complete equilibrium with the temperature of water in the deepest parts of the oceans. Figure 6.3 depicts sea levels predicted to occur between 2081 and 2100 compared to a 1986 to 2005 baseline broken down by the source of the change and four different scenarios that predict the possible climates of the future.

Water can store much more heat energy than air. This fact underlies observations indicating that upward of 90 percent of the earth's energy gain due to climate change is stored in the oceans.[9] Most of this energy gain is found in the 700 meters of water closest to the surface, where the increase in surface water temperature parallels that of the atmosphere. This warmed water becomes less dense and expands, causing it to remain near the surface in spite of upwellings and other factors that promote mixing of deep ocean water with water near the surface. These physical factors have made thermal expansion of the oceans the leading cause of rising sea level.

Because of the physics of heat transfer into the oceans, altering the time course of thermal expansion is extremely difficult.[10] As a consequence,

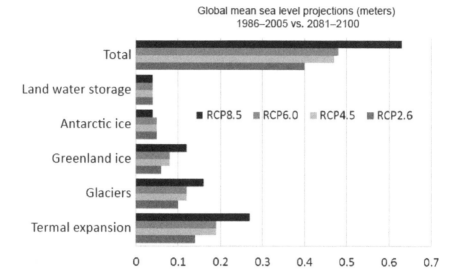

**Figure 6.3**

Future sea levels predicted by the four RCP scenarios. Data extracted from J. A. Church, P. U. Clark, A. Cazenave, et al., "Sea Level Change," in *Climate Change 2013: The Physical Science Basis; Contribution of Working Group I to the Fifth Assessment Report of the Intergovernmental Panel on Climate Change*, ed. T. F. Stocker, D. Qin, G-K. Plattner, et al., 1137–1216 (New York: Cambridge University Press, 2014).

models that predict future sea level increases yield similar results in the near term. This means that even in the extremely unlikely event of a near-instantaneous and sharp reduction in greenhouse gas emissions, sea levels will continue to rise. What's worse, models predict that sea levels will continue to rise due to thermal expansion long after stabilization of the climate. In a rather pessimistic statement, the authors of a recent paper on the reversibility of sea level rise write, "Therefore, despite any aggressive $CO_2$ mitigation, regional sea level change is inevitable."[11]

Huge amounts of water are trapped in the ice that covers Greenland and the Antarctic.[12] The 150-meter deviations from present sea levels that are found in the paleoclimate record are a reflection of changes in these two repositories. The effects of climate change on Greenland ice are becoming increasingly clear: the ice is melting at an accelerating rate and will cause a rise in mean sea level. Antarctic ice is proving to be more difficult to understand and model. As the atmosphere warms, it is capable of holding more water. Some of this water will be deposited as snow in Antarctica, which, in spite of all of the ice that covers it, has near-desert-like conditions because of the relatively small amount of snow that falls each year. Thus, because of warming and the associated increase in Antarctic snowfall, it is likely that there will be an increase in ice in Antarctica. This increase is expected to have major effects during the present century and cause a fall in sea level between 0 and 70 mm.

On the other hand, warming oceans and accelerating movement of Antarctic ice are working in the opposite direction. Research indicates that this trend is most characteristic of West Antarctica. The theoretical concept referred to as *marine ice instability* suggests the possibility of a positive feedback cycle that could result in relatively rapid movement of Antarctic ice from the land surface into the ocean, which would cause a correspondingly large increase in sea level. Although there are two recent independent reports of "unstoppable" movements of Antarctic ice, with one glacier retreating 35 km between 1992 and 2011, a massive collapse of the Antarctic ice sheet does not seem to be likely during the next century or two.[13] However, very long-range projections favor a loss of Antarctic ice resulting in a rise in sea level. A 2015 report with the self-explanatory title "Combustion of Available Fossil Fuel Resources Sufficient to Eliminate the Antarctic Ice Sheet" makes the point that a fifty-eight-meter rise in sea level would follow burning all of the earth's fossil fuels.[14] The good news is that this would take centuries. The bad news is that the most dramatic rise in sea level would occur early on.

Not all human activity has caused sea levels to rise; some has had the opposite effect. The authors of an analysis of the effect of water storage in reservoirs estimated that about 10,800 cubic kilometers of water have been stored on land.[15] This reduction in the amount of water entering the oceans has prevented the oceans from rising by about three centimeters. The number of reservoirs has dropped since around 1980, and the total amount of water being impounded has started to level off, so this effect is likely to be transient.

## Storm Surges and Severe Storms

Storms commonly create a temporary bulge in the water, known as a *storm surge*. Surges, combined with increases in sea level, will make coastal areas more vulnerable to storm-associated flooding. Storms are predicted to increase in intensity as the climate warms.

About 95 percent of the height of a surge is due to the wind as it pushes water ahead of the core of the storm. This effect forms a bulge on the surface of the ocean—seen as a temporary increase in sea level. The low barometric pressure associated with severe storms makes a relatively minor contribution to the developing surge. As the bulge encounters the shallower water near the shore, it becomes more pronounced. Certain shoreline configurations may act as a funnel, leading to further increases. These surges are superimposed on the more easily predicted astronomical tides caused by the moon. The storm surge plus the astronomical tide is referred to as a *storm tide*. Storm tides are most damaging when the landfall of the storm surge coincides with an astronomical high tide. Waves that are superimposed on the storm tide add an extra measure to the increase in sea level.

Superstorm Sandy, which made US landfall on October 29, 2012, was one of most damaging storms ever to hit the United States. It was an atypical event. It occurred relatively late in the hurricane season and had a highly unusual path. Sandy made an abrupt turn toward the west, striking the mid-Atlantic shore, rather than following a more typical path to the northeast. As it evolved from a tropical hurricane deriving its energy from the ocean to a post-tropical storm that derived its energy from the atmosphere, Sandy grew to an immense size. As a result, even though it was only a category 2 storm when it made landfall, the associated storm surge of 8.99 feet at Battery Park in lower Manhattan was more than twice as high as its nearest rival.[16] The eighteen-inch rise of sea level that has occurred since 1850 combined with the fact that the surge coincided with a typical astronomical high tide resulted in a water level that was over thirteen feet above the mean low-level water mark. The results were catastrophic.

The New York subway system floods when sea level rises to 10.5 feet above the mean low-level water-reference point, and so the storm surge of just over thirteen feet flooded the system.[17] Many communication centers were also engulfed. Other critical infrastructure elements, such as Bellevue Hospital, also flooded. When the final toll was counted, at least 233 people were believed to have died and property damage was estimated to be about $68 billion.[18]

This disaster pales by comparison to Cyclone Bhola, which struck what is now Bangladesh on November 12, 1970.[19] This cyclone (i.e., a hurricane occurring in the South Pacific or Indian Ocean) grew rapidly in the Bay of Bengal. It slowed as it approached the Ganges-Brahmaputra Delta, which allowed the storm surge to grow. Because the northern part of the bay narrows, acting as a funnel, the surge gained still more height. Silt from the rivers blocked backflow from the surge, compounding the problem. The surge eventually grew to an estimated twenty-five feet, converting the low-lying coastal region into what was described as a "death trap." Sustained winds of 130 mph created waves that added to the death and destruction. The death toll was never determined with accuracy, but estimates ranged up to five hundred thousand, making it one of the worst natural disasters of the century.

As discussed in chapter 1, political stability and stakeholder involvement are among the ideal prerequisites for effective management of the elements of climate change. The area struck by Cyclone Bhola, which was then part of East Pakistan, was a country where a disaster was waiting to happen. The cyclone triggered a substantial amount of political unrest in a region that already faced immense problems. The ensuing Bangladesh Liberation War led to a change of government and the eventual establishment of Bangladesh as an independent nation in 1971.

Thus, it should not have been too surprising that there was virtually no meaningful response to this cyclone: shallow water made the area inaccessible to ships. There were virtually no helicopters available to come to the aid of the stricken population. Although the United States had many helicopters in Vietnam, those that eventually arrived came from the continental United States.

The 1970 Bhola disaster was, if nothing else, a powerful learning experience. Analyses of the catastrophe led to substantial improvements in disaster response planning and public health preparedness in Bangladesh.[20] High-tech improvements included the creation of early warning systems. Low-tech measures allowed these warnings to be distributed by volunteers riding bicycles. As a result, when a cyclone of similar intensity made landfall in the same area in 1991, there were about 140,000 deaths: still a lot, but nowhere near the total from the 1970 storm and floods. Additional improvements reduced the death toll to 4,234 when Cyclone Sidr, a category 5 storm with winds of 160 mph, ravaged the area in

2007.[21] This decline in the death toll shows that adaptive disaster preparedness measures can make an enormous difference.

## Predicted Flooding

In a *New York Times* feature titled "What Could Disappear," journalists published a series of maps showing the extent of flooding that would be expected given various scenarios.[22] The authors used elevation maps from the US Geological Survey and tidal data from the National Oceanographic and Atmospheric Administration to calculate the amount and location of flooding expected if sea level rose five, twelve, or twenty-five feet, as envisioned by the authors of the IPCC Fourth Assessment Report. With a rise of five feet, flooding would engulf 26 percent of Cambridge, Massachusetts, 19 percent of Charleston, South Carolina, and 20 percent of Miami and 94 percent of Miami Beach, Florida. Sacramento, California, thought of as being "inland," would experience 4 percent flooding. These estimates do not include any of the additional increases in sea level from storm surges or just plain bad luck associated with storm surges that would coincide with unusual but predictable high tides. Additional details from that report are shown in table 6.1.

The data in the table show the extent of flooding without any attempt to monetize the result or to evaluate the impact on affected populations. However, it is not difficult to imagine that there would be an enormous impact from a five-foot increase in sea level on cities like Miami, Miami Beach, and New Orleans. An increase in sea level of that magnitude would engulf major portions of these important metropolitan areas. One prediction for Miami, Florida, suggests that a large storm surge could cause damages measured in the tens of billions of dollars.[23]

Port cities around the world are highly vulnerable to the effects of increasing sea level. A 2005 analysis of ports with more than one million inhabitants concluded that approximately forty million people are currently vulnerable to coastal flooding due to the combined effects of sea level increases and storm surges that would be expected during a one-hundred-year event, one that has a 1 percent probability of occurring in any given year.).[24] The value of exposed assets was estimated to be around $3 trillion, or 5 percent of the 2005 global gross domestic product (GDP). The United States, Japan, and the Netherlands have the greatest financial

Table 6.1

Sea level rise and percent of expected flooding in major US cities

| City | Percent flooding, 5 ft. rise | Percent flooding, 12 ft. rise | Percent flooding, 25 ft. rise |
|---|---|---|---|
| New York City, NY | 7 | 22 | 39 |
| Boston, MA | 9 | 24 | 37 |
| Cambridge, MA | 26 | 51 | 86 |
| Jersey City, NJ | 20 | 46 | 62 |
| Miami, FL | 20 | 73 | 99 |
| Miami Beach, FL | 94 | 100 | 100 |
| Mobile, AL | 4 | 19 | 36 |
| New Orleans, LA | 88 | 98 | 100 |
| San Francisco, CA | 6 | 11 | 19 |
| Seattle, WA | 4 | 9 | 13 |
| St. Petersburg, FL | 32 | 49 | 70 |
| Tampa, FL | 18 | 32 | 50 |
| Washington, DC | 2 | 7 | 14 |

*Notes:* The authors did not provide a rationale for choosing a five-foot increase, but that is well within the boundaries expected from climate change scenarios presented in the IPCC Fifth Assessment Report. The twelve-foot increase was chosen because it was deemed likely by 2300 as long as there are only moderate reductions in greenhouse gas emissions. The extreme twenty-five-foot increase in the more distant future was chosen on the basis of historical climate data that have shown changes in sea level of about 150 feet at a time when there was no polar or Greenland ice. For details, see B. Copeland, J. Keller, and B. Marsh, "What Could Disappear," *New York Times*, November 24, 2012.

exposure. By 2070, the worldwide population that would be ravaged by a one-hundred-year event could grow by a factor of three due to increases in sea level, coastal subsidence, population growth, and urbanization. At that time, total asset exposure could rise to as much as 9 percent of the global GDP. The study's authors conclude that there are significant potential benefits associated with protecting cities to reduce risk.

In 2011, researchers published an extremely detailed analysis of the effects of sea level rise on the city of Copenhagen, Denmark.[25] The researchers report that with a relatively modest increase in sea level of 0.5 to 1.0 meters the total insured value for property at risk would be nearly €2.3 billion, including around €1 billion for residential property, €900 million for commercial property, and €400 million for industrial property. After a hypothetical 0.5-meter rise in sea level and a 1.5-meter storm surge, the researchers predict that around 3,500 jobs would be lost in the personal services sector three months after the event. They estimate that this would fall to about 1,300 missing jobs one year after the flooding. Construction jobs would increase by about two hundred at the three-month interval and by about 1,800 after one year. Full recovery would take many years. Greater increases in sea level would lead to even larger losses, disrupting the economy of the city and the nation. Surrounding the city with a coastal flood-protection system consisting of dikes and sea walls was suggested to be a relatively simple task, with an estimated cost of several hundred million euros. This system would require a budget to fund maintenance, pumps, drainage systems, and other infrastructure improvements. The researchers concluded that such a system would be a rational investment.

### Storms of the Future

Predicting future tropical hurricane and cyclone activity is one of the most difficult tasks facing meteorological researchers. One might expect that the anthropogenic increases in land surface and sea surface temperatures and associated increases in atmospheric water content would set the stage for increases in the number and intensity of destructive storms. However compelling this hypothesis might be, support is elusive. Some of the difficulties in this realm center on a lack of highly accurate information concerning past storms. In addition, hurricanes and cyclones are somewhat rare events. Although these storms are most likely to occur at specific times of the year, the variance in annual occurrence data makes it difficult to define trends. This fact is illustrated by an analysis of damage due to hurricanes between 1900 and 2005.[26] In many years, there was virtually no damage, whereas in others—exemplified by the 1926 data—normalized damages were over $150 billion. This degree of variance makes it difficult, if not impossible, to detect trends.

However, there are some useful long-term data—such as an example derived from an examination of sediments retrieved from a low-lying lake near Boston, Massachussetts.[27] Core samples from the lake were correlated with known hurricanes documented in the written records from the era. An examination of ten out of eleven candidate layers of sediment showed a concordance between their content and known category 2 and

3 storms. This established a basis for examining older, deeper layers that extended back one thousand years. From this examination, study authors concluded that hurricane activity was high between the twelfth and sixteenth centuries, and lower during the eleventh century and between the seventeenth and nineteenth centuries. Although they are important, these data are insufficient for forecasting future storm activity.

In the current era of satellite-aided meteorological research, a much more accurate inventory of storms is possible. However, satellite observations do not cover all of the earth evenly, and some deficiencies in the data gathered in the present still exist.

In spite of these limitations, remarkable progress toward accurate prediction of future hurricane activity has been made in the past several years. Converging results predict an increase in the number of category 4 and 5 storms. One such study predicts a reduction in the total number of hurricanes but also that the number of high-intensity storms will double by the end of the twenty-first century.[28] They forecast that the largest increase will take place in the western Atlantic Ocean north of the line at twenty degrees north latitude. The authors of another review, published at about the same time, came to similar but somewhat more specific conclusions in their analysis.[29] They predict that anthropogenic increases in greenhouse gases will lead to a shift toward more intense storms of between 2 and 11 percent by the end of this century. They further predict that the intensity of these stronger storms, as measured by precipitation rates within 100 km of the storm's center, will increase by about 20 percent. Finally, they report that multiple models predict a reduction in the frequency of tropical storms of between 6 and 34 percent by 2100.

## The Trajectory toward the Future

A large portion of the earth's population is at risk as sea levels rise, storms increase in their intensity, and people flock to already crowded, urbanized coastal areas. In the summer of 2015, the news was filled with descriptions of hundreds of thousands of refugees struggling to leave Africa and the Middle East, hoping to find a better life in the nations of the European Union. There is real potential that these numbers and the suffering they represent could be dwarfed by the number of refugees created by climate change–associated coastal flooding. As always, the most vulnerable will bear the heaviest burden.

Predicting the future is always difficult. Here, the task is made difficult because of incomplete data from the past and the complexities associated with looking toward the future. However, it is clear that the climate is warming and too little is being done to mitigate and adapt to the changes that seem too likely to occur. Even if a miracle were to occur that halted emissions of greenhouse gases overnight, the sea would continue to rise. Rising temperatures will evaporate more water from the warmer oceans, fueling more powerful storms. We may be in for a rough ride.

# 7

## Air Pollution, Air Quality, and Climate Change

We need to know more about the total environment ... only by reorganizing our Federal efforts can we develop that knowledge, and effectively ensure the protection, development and enhancement of the total environment itself.
—Richard Nixon, message to Congress establishing the EPA, July 9, 1970

After World War II, Donora, Pennsylvania, was a prosperous town of about fourteen thousand inhabitants located on a bend of the Monongahela River. There were many well-paying jobs in the town's major industry, a huge complex consisting of open hearth and blast furnaces at one end and a zinc works at the other. Coal was plentiful in Pennsylvania. Coal and iron ore went in one end of the linear array of factories, and galvanized wire, nails, and other products emerged from the other end. Executives lived in plush homes on the hillside that overlooked the zinc works. Others lived in Cement Town, at the other end of the city that bristled with smoke stacks, viewed as a sign of prosperity. Nobody paid much attention to the fact that virtually all of the plants on both sides of the river had died.

Things changed dramatically and suddenly on Thursday, October 28, 1948, when a temperature inversion trapped the gases and particulates emitted by the mill. The smog was so dense that residents walking down the street could no longer see the buildings that were familiar landmarks. Then, people began to become sick and die.

Firefighters struggled to bring oxygen tanks to the afflicted. The Board of Health set up an emergency aid station and temporary morgue in the community center. The mill management refused to believe that the pollutants spewing from the factory were responsible for the crisis. After all, they said, the mill was doing what it had been doing for thirty years.

The exact death toll will probably never be known with certainty, but at least twenty-one died and hundreds were sickened. Investigations that followed linked the disaster to the oxides of sulfur, fluorine, and other unchecked emissions from the zinc works and the rest of the factory complex.

Now, if you walk down McKean Avenue past abandoned stores, it is hard to find traces of the mill. However, at the corner of McKean and Sixth Street, you will find one of the few painted storefronts: the Donora Smog Museum, located in what used to be a Chinese restaurant (http://bit.ly/1MAROOV, accessed September 28, 2015). On a lucky day, you will meet Brian Charlton, one of the town's schoolteachers and the museum's curator. He will tell you about the killer smog, the subsequent investigations, and how it all led to the town's new slogan, "Clean Air Started Here." Or he may regale you with tales of Stan "the Man" Musial, who was from Donora and whose artifacts occupy a share of the museum.

The Donora disaster was precipitated by a weather event (the formation of an inversion layer)—not climate change—and by uncontrolled emissions from the Donora mills. Its importance lies far beyond the toll on the residents of the town: this event marked the beginning of efforts to protect health by controlling hazardous air pollutants. In its analysis of the benefits of these protections, the EPA concentrates on two: particulate matter and ozone. This chapter will focus on both protections, with emphasis on ozone and how the EPA makes its decisions on air quality standards.

### A Brief History of Air Pollution Controls

The Donora disaster was not the first hint that smoke is bad for health. One of the earliest recorded warnings about poor air quality was published in 1661, when John Evelyn warned "His Sacred Majestie" of "The Inconvenience of the Aer and Smoak of London" in the treatise he submitted to his king and Parliament.[1] A number of more contemporary events cemented this link, including accounts of the infamous 1952 London "killer fog" that caused an estimated 12,000 deaths.[2]

Donorans will tell you that their tragedy led the US government to focus on the relationship between air quality and health. Multiple investigations that followed the Donora tragedy began to coalesce in 1955, when Congress passed the Air Pollution Control Act, which funded research into air pollution. The idea of controlling air pollution came into prominence in 1963 with the passage of the Clean Air Act (CAA), which provided funds for the development of air pollution monitoring and control under the auspices of the US Public Health Service. Actual controls on air pollutants were enacted seven years later.

The quote that introduces this chapter is from President Richard Nixon's 1970 executive order that established the Environmental Protection Agency. With broad bipartisan support in Congress, the EPA was created by law, and the CAA was amended to give the new agency the authority to regulate air pollutants. The amended act required the EPA to establish National Ambient Air Quality Standards (NAAQS) for the most offensive pollutants, of which there are six, referred to together as the *criteria pollutants*: lead, carbon monoxide, nitrogen dioxide, sulfur dioxide, ozone, and particle pollution. For each, there are two standards. The first is referred to as the *primary standard* and is designed to set a concentration limit with an "adequate margin of safety," thereby enabling the EPA to fulfill its mission, "to protect human health and the environment." The primary standard makes special reference to "sensitive" populations, such as children, asthmatics, and the elderly. The *secondary standard* is designed to protect the general welfare and focuses on preventing reductions in visibility, damage to crops and other forms of vegetation, animals, and buildings. Portions of the country that meet the standards are referred to as *attainment areas*. Areas that fail to meet the standard are referred to as *nonattainment areas*.

The removal of lead, a potent neurotoxin, from gasoline was one of the earliest and most important accomplishments of the CAA. Although this was controversial at the time, we now take lead-free gasoline for granted. The action led to a dramatic drop in the lead burden of children and an increase in the average IQ of over five points per child.[3]

The CAA was amended again in 1990, during the administration of George H. W. Bush. This amendment also passed with broad bipartisan support in the House of Representatives (401 to 21) and the Senate (89 to 11). This is the so-called acid rain amendment. Although the focus of the amendment was on the authority to regulate the sources of oxides of nitrogen ($NO_x$) and sulfur ($SO_x$) that were causing acid rain and damaging lakes and forests, it also enabled the EPA to regulate $NO_x$ and $SO_x$ along with chemicals that were depleting the ozone layer—most notably, the chlorofluorocarbons.

Under the authority of the CAA, the EPA has promulgated and amended a variety of regulations that establish NAAQS for each of the criteria pollutants. The CAA also requires the EPA to publish periodic reports to Congress that quantify the CAA's costs and benefits. These reports are one of the most important sources of data that make clear-cut links between improvements in air quality and subsequent savings in the total costs associated with poor health. In the most recent of these reports, the EPA predicts that by 2020 the CAA will result in nearly $2 trillion in health benefits each year, at an annual cost to industry of $65 billion.[4] These benefits are due largely to reductions in the atmospheric concentrations of particulate matter and ozone, two criteria pollutants. An account of the deaths prevented, asthma attacks averted, and other health benefits attributable to the CAA and the anticipated new ozone standards are presented in table 7.1 and table 7.2. The importance of these pollutants provides the rationale for the focus of this chapter and why the

Table 7.1
Annual health benefits attributable to the Clean Air Act

| Health effect reduction | Pollutant | Year 2020 |
|---|---|---|
| Adult PM2.5 deaths | PM | 230,000 |
| Infant PM2.5 deaths | PM | 280 |
| Ozone deaths | Ozone | 7,100 |
| Chronic bronchitis | PM | 75,000 |
| Acute bronchitis | PM | 180,000 |
| Heart attack | PM | 200,000 |
| Asthma exacerbation | PM | 2,400,000 |
| Hospital admissions | PM, Ozone | 135,000 |
| Emergency room visits | PM, Ozone | 120,000 |
| Restricted activity days | PM, Ozone | 110,000,000 |
| Lost school days | Ozone | 5,400,000 |
| Lost work days | PM | 17,000,000 |

Source: US Environmental Protection Agency, Office of Air and Radiation, *The Benefits and Costs of the Clean Air Act from 1990 to 2020* (Washington, DC: EPA, 2011).

**Table 7.2**
Annual health benefits expected after new ozone air quality standard, year 2025; includes ozone and particle-reduction effects

| Disease category | Range of expected effects |
| --- | --- |
| Premature deaths | 710–4,300 |
| Asthma attacks, children | 320,000–960,000 |
| Missed days of school—children | 330,000–1,000,000 |
| Missed work days—adults | 65,000–180,000 |
| Emergency room visits—asthma | 1,400–4,300 |
| Acute bronchitis—children | 790–2,300 |

*Source:* US Environmental Protection Agency, "Regulatory Actions: Ground-Level Ozone," EPA, December 10, 2014, http://www3.epa.gov/ttn/naaqs/standards/ozone/s_o3_index.html.

Intergovernmental Panel on Climate Change includes air quality as a climate change health target (for more details, see figure 10.1).

## Ozone

*Ozone* ($O_3$) is a pale blue gas. It has a distinctive pungent odor similar to that of chlorine. It is formed in the lowest layer of the atmosphere, the *troposphere*, which extends from the surface of the earth to an altitude of about eleven miles. The troposphere is thicker near the equator and thinner at the poles. In the troposphere, ozone is an important constituent of ground-level smog. Ozone is also formed in the next-highest layer, the *stratosphere*, which extends upward above the troposphere to an altitude of just over thirty miles at the equator and to just over eight miles at the poles. In the stratosphere, ozone protects us from the ultraviolet rays of the sun that cause skin cancer and damage crops and other plants. Hence the phrase, "Ozone: good up high, bad nearby."

Ozone is both formed and destroyed in the stratosphere as the result of the actions of ultraviolet (UV) light.[5] UV light from the sun splits molecular oxygen ($O_2$) into two atoms of oxygen (O) in the middle portion of the stratosphere. This highly reactive form of oxygen combines rapidly with $O_2$ to form ozone ($O_3$). In upper layers of the stratosphere, high-energy UV rays from the sun attack ozone to form $O_2$ and O, thus creating a cycle of formation and degradation. The heat generated by these reactions and the associated trapping of the energy from the sunlight that drives them warms the upper portions of the stratosphere.

The chemical reactions involving tropospheric ozone are much more complex. The basic reactions involve the criteria pollutants nitric oxide and carbon monoxide along with volatile organic compounds (VOCs) in the presence of heat, sunlight, and diatomic or regular oxygen. Methane is one of the VOCs involved in these reactions. Because of the importance of methane as a global warming gas, VOCs are commonly subdivided into (1) methane; and (2) everything else, or nonmethane VOCs (NMVOCs). Isoprene is the most important of the naturally occurring NMVOCs, with estimated annual emissions of between four and six hundred million metric tons per year, an amount that is thought to be similar to natural emissions of methane.[6] Isoprene concentrations are likely to increase due to more exuberant plant growth stimulated by increases in the atmospheric carbon dioxide concentration and warming due to climate change. Many NMVOCs are created by human activity. Some come from petroleum products, whereas others come from the transportation industry, paint, and industrial chemicals. Uncertainties in the prediction of future emissions of VOCs contribute to problems in forecasting future ozone levels. Finally, some ozone is formed from oxides of nitrogen in bolts of lightning. This mechanism is also likely to become more important as climate change causes an increase in the number of severe storms.

Ozone is removed from the troposphere by several chemical reactions. The ozone–water vapor reaction is critical and is expected to help keep ozone levels from rising in humid southern portions of the United States To a large degree, the ozone–water reaction is expected to balance the effects of an increase in temperature that would otherwise lead to an increase in the concentration of ozone.

Tropospheric ozone is also a short-lived greenhouse gas. It is responsible for about 15 percent of the increase in the greenhouse effect that has taken place since the beginning of the industrial age and the onset of our warming climate. From a technical perspective, tropospheric ozone accounts for about $0.35$ W/m$^2$ in its role as an RF agent, a quantitative measure of the greenhouse effect (see chapter 2 for more details). To place this in a broader perspective, total RF has increased by about $2.4$ W/m$^2$

since the onset of the industrial age, according to the most recent IPCC report (see also chapter 2).[7]

Ozone is a powerful oxidizing agent. It combines chemically with a wide variety of other molecules. In our bodies, these chemical reactions damage tissues—particularly in the respiratory system, where they cause inflammation and trigger immunological reactions. These responses are the link between ground-level ozone and asthma, one of the most prevalent diseases in the United States. The CDC estimates that 7.3 percent of adults and 8.0 percent of children suffer from asthma.[8]

Because of the importance of ground-level ozone, its concentration is monitored closely using a variety of ground- and satellite-based techniques. The EPA uses these ozone concentration data to compute an air quality index (AQI) that provides health warnings and guidance to individuals and communities. Interested parties can access the AQI at the EPA's Enviroflash website (www.enviroflash.info), and the AQI is commonly included in weather forecasts. The AQI ranges from 0 to 500, where 0–50 means good air with no health concerns, 51–100 indicates moderate health concerns, 101–150 indicates unhealthy air for sensitive groups, 151–200 indicates unhealthy air, and 201–300 and 301–500 indicate very unhealthy and hazardous air, respectively.

**Atmospheric Ozone Concentration Standards: Past, Present, and Future**
This is a story that goes beyond the realm of health and climate change to include history, science, politics, and the law. Like oil and water, these elements do not always mix well.

Emission controls focused largely on the criteria pollutants have improved the quality of the air in the United States. Ozone levels are 33 percent lower now than they were in 1980, and 90 percent of the areas that failed to meet ozone air quality standards in 1988 comply with the 75 parts per billion (ppb) standard.[9] The situation is similar in Europe, where there have also been air quality improvements. However, ozone levels have more than doubled in Asia as a direct result of the economic expansions in that region in the absence of controls on air emissions (see section 2.2.2.4 of the IPCC Fifth Assessment Report for details).[10]

The primary, or health-based, air-quality standard for ozone was set at 75 ppb in 2008 (i.e., there are seventy-five ozone molecules in every billion air molecules). The EPA's language that defines standards is often convoluted—and this is particularly true for ozone. As of December 28, 2015, the primary standard was set as the "annual fourth-highest daily maximum 8-hour concentration, averaged over 3 years."[11] Translated, this means that the average ozone concentration in any given eight-hour time interval may not be greater than 75 ppb on more than three occasions during any given three-year period. Understandably, this is usually shortened to "the 8-hour standard." As required by the CAA and in order to update required health protections, the EPA began a review of the standard in the fall of 2014. The intent was to lower the standard to somewhere between 60 ppb and 70 ppb. Based on 2011–2013 monitoring data, 358 counties—including most major metropolitan areas in the United States—would have been noncompliant if a 70 ppb standard had been in effect. If the standard were 65 ppb, a total of 558 counties would be in violation and classified as nonattainment areas.[12] California is excluded from these data because the Air Resources Board has set their eight-hour standard at 70 ppb.

The path to the 2014 reevaluation is complicated and heavily influenced by political considerations. In 1971, the very first primary standard was 80 ppb, averaged over one hour and not to be exceeded by more than one hour per year. The standard was revised in 1997, when it was lowered to 80 ppb averaged over eight hours. In 2008, the eight-hour standard was lowered to 75 ppb, where it remained until 2015. This is where the story becomes complex as political considerations emerge.

When the EPA lowered the eight-hour standard to 75 ppb, it did so against the unanimous advice of its own Clean Air Scientific Advisory Committee (CASAC), which had urged setting the standard to between 60 ppb and 70 ppb. This deviation was deemed a political decision made by the George W. Bush administration. The standard was revisited again from 2009 to 2010 during the administration of Barack Obama. However, on a Friday afternoon, a time apparently chosen to minimize media scrutiny, the EPA announced that it had suspended plans to revise the standard in spite of the formal publication of a proposed rule in the Federal Register and the receipt of numerous comments from the public and other stakeholders. Again, this was viewed as a political decision. This time, it seemed to be designed to take political heat off the EPA and to shield candidates during the forthcoming election cycle. On the

Wednesday prior to the US Thanksgiving holiday in 2014, the EPA announced that it was once again reconsidering the standard and opened a new round of comments on a proposed rule.

Lawsuits ensued, and an October 1, 2015, deadline to publish a final standard was imposed via a consent decree. The new standard was set at 70 ppb. After intense lobbying, industry leaders complained, advocating leaving the standard at 75 ppb. Public health groups and environmentalists that had advocated for a more stringent rule with a standard of 60 ppb argued that the EPA had failed to protect public health adequately. The new standard does not take effect immediately. Nonattainment areas have until 2020 or late 2037 to meet the new standard. By then, the EPA will be required to reevaluate the standard. More lawsuits seem likely.

### The Science behind the Standards

So what is the evidence that concentrations of ozone higher than the 75 ppb standard are bad for health? During the course of my medical career, I reviewed many articles for a variety of scientific and medical journals as part of the peer-review process, and for many years I was a member of my hospital's science review and ethics committees, including three years as chair of the ethics committee, formally known as the institutional review board (IRB). The primary review considerations are related to methodology, results, and conclusions—that is, whether the science is sound. However, there is more to consider, such as the ethics involved. Usually this means that there is a declaration that an independent review committee, such as the one I chaired, has approved the research. It is also important to ask whether potential issues related to conflicts of interest and bias have been managed properly, which may involve determining the source of the funds used to sponsor the study. Therefore, when I reviewed the human inhalation studies, I saw them from my perspective as a university professor who worked in a medical environment in which I taught a course in evidence-based medicine and chaired the research ethics committee in my hospital.

Because the route for exposure is via the lungs and numerous studies have linked ozone exposure to exacerbations of asthma and other respiratory conditions, several ozone-inhalation studies in normal volunteers have been published. The authors stated that federal and international standards for the ethical conduct of research involving human participants were met and informed consent had been obtained. Importantly, the brief administration of ozone to healthy individuals was thought to pose little risk. Unhealthy individuals and children were excluded. These studies appear to have had two objectives: to better understand the effects of ozone on the human respiratory system and to influence the EPA as it seeks to update the air quality standard for ozone.

In its proposed rule, the EPA appears to rely on many of the studies that I have reviewed. Some were funded in part by the American Petroleum Institute, an industry-supported group.[13] This fact creates the potential for bias even though the results were published in peer-reviewed scientific literature. It is virtually certain that research participants were paid. This is permissible, but setting an appropriate level for reimbursement is tricky. The ethics committee must decide that the amount does not create the possibility for coercion. Coercively high payments discourage dropouts from the study and underreporting of adverse effects that could result in dismissal from the study and a loss of income. Both have the possible effect of introducing bias into the results. The study volunteers were healthy men and women free of respiratory problems who had not been exposed to high ozone concentrations. Some were competitive athletes; those who were not were commonly regular exercisers. Thus the research participants were healthier than typical Americans. This creates yet another opportunity for bias when research results are extrapolated to the American population as a whole.

Participants were exposed to ozone concentrations that ranged from 40 ppb to 120 ppb during several 6.6-hour intervals. Participants exercised during exposures and underwent basic pulmonary function tests. Participants also completed a questionnaire known as the total symptom score (TSS) inventory as a part of the evaluation. Although some of the inhalation protocols and statistical analyses seemed somewhat contorted in a way that minimized explaining the effects of the ozone inhalation, a close evaluation of the data shows that virtually all studies found effects on the pulmonary function tests, the TSS, or both at relatively low $O_3$ concentrations. A reevaluation of one of these studies focused on just one of the many inhalation protocols and found clear effects of 60 ppb ozone on breathing function.[14]

Another study I examined was conducted by EPA scientists, reported in 2011, and again evaluated young healthy adults who were exposed to 60 ppb of $O_3$.[15] In addition to significant reductions in two standard breathing tests, these investigators found evidence for inflammation in the airway by studying white blood cells (polymorphonuclear leukocytes) in the sputum sixteen to eighteen hours *after* $O_3$ exposure. Because ozone is a potent oxidizing agent, investigators also measured a gene called glutathione S-transferase mu 1 that affects responses to oxidative stress. Participants who carried this gene experienced significant effects on one of the breathing tests. They concluded that inhaling $O_3$ at a concentration of 60 ppb affects lung function and triggers an immunological response. These are hallmarks of asthma.

In a 2015 report, a Taiwanese group measured pulmonary function in a group of almost 1,500 nonasthmatic children.[16] They performed typical tests of pulmonary function and correlated these with ozone data from monitoring sites near the children's schools. The average ozone concentration during the monitoring period was 29 ppb. The maximum never exceeded 58 ppb. They found that on a day after an ozone peak there was a reduction in lung function as measured by a standard breathing test. Younger children, six to ten years of age, were more affected than older children.

When new $O_3$ concentration limits are set, it is important to consider additional factors that do not always show up in the statistical analyses found in these reports. It seems likely that individuals with asthma, other pulmonary problems, or other chronic diseases, especially of the cardiovascular system, might be more sensitive to ozone than the super-healthy adult research participants were. It is possible to make a convincing argument that the research participants in the published studies were not "normal" subjects but were in fact healthier than normal and part of a population that is the *least* likely to sustain an adverse effect after the inhalation of $O_3$. Children are also likely to be more susceptible to ozone at the concentrations studied, as suggested by the observations made on Taiwanese children, because each minute they breathe more liters of air per unit weight than adults. This means that the dose of ozone at the same ozone concentration is higher in children. The dose differs from the concentration and may be a more important measure. Dose is determined by the concentration of ozone in the air multiplied by the amount of air inhaled and time. Because children and other sensitive populations were excluded from the studies, their results may have only limited applicability to the EPA's task, which is to protect "sensitive populations ... with an adequate margin of safety."

## Future Ozone Concentrations

Global and regional modeling efforts are needed to predict the effects of climate change on future atmospheric ozone concentrations. The variables that are the most important determinants of the future ozone concentration include the temperature itself, which is likely to increase as the result of climate change; the atmospheric concentration of ozone precursors, including nitrogen oxides and natural and anthropogenic VOC concentrations, some of which will increase due to climate change; ozone produced by lightning, which may also increase in prevalence; air currents that mix stratospheric and tropospheric ozone; and the concentration of water vapor, a factor that is also likely to be altered by climate change. Water vapor is the only one of these variables that leads to a decrease in the ozone concentration.

There are typically two components to global ozone models: a global circulation model (GCM) and a chemical transport model (CTM). GCMs simulate atmospheric climatic phenomena, such as heat flow, whereas CTMs simulate the behavior of a given chemical species, such as VOCs. This GCM-CTM combination yields data that predict worldwide background ozone concentrations. Additional modeling efforts, based on local assumptions concerning emission scenarios, produce more regional data that are used to predict ozone concentrations in a given city.

The authors of a relatively recent review of twelve studies that compared data from the year 2000 with what they expected in 2050 found three investigations that predicted an increase in the global ozone concentration and nine that predicted a decrease.[17] Projected changes ranged from a 12 percent decrease to a 5 percent increase. These studies focused on what the EPA terms *policy-relevant background ozone concentrations*, or ozone that would be found in the surface layer of the troposphere over the United States *in the absence* of additional North American anthropogenic emissions. Differences in assumptions, particularly the effects of lightning and emissions of isoprene (both of which are likely to increase due to climate change), account for a great deal of the variability among studies. The review's authors conducted their own study, utilizing an IPCC scenario that assumes a balance between fossil fuel and other energy sources. Using the IPCC A1B scenario, they projected a 2 percent increase in the global ozone burden. In the eastern United States, rising emissions of ozone precursors that would drive ozone concentrations up were canceled out by increases in the tropospheric water vapor content, which removes ozone from the air, in spite of increases in temperature. In the western part of the country, which is drier, they project an increase of between 2 and 5 ppb in the ozone concentration.

To begin to address the issues posed by climate change related to ozone levels in US cities, the authors of another recent study used models based on an IPCC scenario that predicts a continued high rate of $CO_2$ emissions (the IPCC A2 scenario).[18] This scenario predicts temperature increases of between 1.6°C and 3.6°C by 2050 relative to 1990 temperatures. Because of difficulties predicting anthropogenic ozone precursor emissions, these authors assumed that such emissions would remain constant in the future. Presumably, population increases and reduced per capita emissions were expected to offset each other. Emissions of natural ozone precursors were included in the authors' modeling effort. They predicted that there would be an increase in the summertime one-hour ozone concentration of 4.8 ppb on average for the cities studied. The largest increase was 9.6 ppb. Cities with an increase of 12 percent or more included Chicago, Illinois, Cleveland, Ohio, Columbus, Ohio, and Detroit, Michigan. The authors predicted a 68 percent increase in the number of days each summer for which the ozone concentration would exceed the present eight-hour standard of 75 ppb. Although they considered the entire United States, the cities with increases were all in the Mississippi River Valley and the eastern half of the country. Because the results of ozone modeling are highly dependent on model assumptions, other studies reached different conclusions. In one of these studies, in which investigators used models that predicted an emissions decrease of more than 50 percent in oxides of nitrogen and a balanced use of fossil and nonfossil fuels, the investigators found that there would be a reduction in future ozone concentrations that ranged between 11 and 28 percent in the eight-hour ozone concentration in different regions of the United States.[19]

The summer of 2015 set new records for the number and severity of wildfires as a consequence of the severe drought that affected significant parts of the nation. These fires produce particulate matter (PM) and are likely to affect the concentration of ozone. The summer of 2004 was also a bad year for wildfires in North America. Significant portions of the forests in Alaska and the Yukon Territory were consumed by fires. The smoke, carbon monoxide, and other emissions drifted toward the northeastern part of the United States and Canada. To assess the impact of these fires on ozone levels, researchers employed data from aircraft-based measuring devices to estimate the additional carbon monoxide burden in the troposphere over New England that arose from these fires.[20] The researchers found that around 30 percent of the carbon monoxide in the atmosphere, a criteria pollutant and an ozone precursor, could be attributed to these distant fires. In a communication with colleagues, the study authors learned that this pollution plume was also detected in Europe. It is likely that carbon monoxide plumes from 2015's fires will affect the eastern part of the United States as they did in 2004.

From these and other studies, it seems possible—even likely—that uncontrolled emissions of greenhouse gases and ozone precursors will lead to increases in the ozone concentration in the most highly polluted portions of the country. If these increases occur, they will have important implications for the health of the exposed populations. One take-home message is clear; there are at least two ways that increases in the ground-level ozone concentration can be prevented: (1) mitigate climate change through effective controls on greenhouse gas emissions to prevent increases in surface temperatures that drive the chemical reactions that produce ozone and (2) control anthropogenic emissions of ozone precursors, particularly methane.

### Is There a Climate Change Penalty?
It is likely that the effects of climate change will lead to increases in the ozone concentration in some parts of the world, particularly in areas with high levels of air pollution due to high concentrations of VOCs and/or low levels of atmospheric water vapor, such as the Southwest. To be certain that the quality of the air at a future date meets the new standard, the initial target concentration must be lower than any new standard. An example may help to explain this point: if the new standard is set at 60 ppb and air quality modeling predicts that the ozone concentration will

rise by 5 ppb in five years due to hotter weather induced by climate change, then the concentration five years after the standard is finalized will be 65 ppb and the region will be noncompliant. To remain compliant during the five-year period, the new target should be 55 ppb, so that when the predicted 5 ppb increase due to climate change is added, the ozone concentration will be 60 ppb and thus in compliance. The 5 ppb offset is referred to as the *climate change penalty*.[21] Research by at least one group predicts that we should include a climate penalty when planning for the future.[22] Of course, the additional reductions undertaken to anticipate a concentration rise would have health benefits, so perhaps the term *penalty* is not appropriate. A *climate change protection* might be a more appropriate and accurate term.

## Particulate Matter

Particulate matter is one of the deadliest of all of the pollutants in the atmosphere. In addition, it is the other criteria pollutant that will be affected by climate change. PM is not a single entity but a category that includes objects with a variety of sizes, shapes, and, importantly, compositions. As science and epidemiology have progressed, it has become evident that PM is a major a risk factor for the development of a variety of diseases.[23]

Atmospheric modeling studies predict that there will be changes in the PM concentration due to the mixing of atmospheric layers and changes in the movement of air from tropical toward polar latitudes. These changes are likely to be more pronounced for PM than for ozone for two reasons: (1) the region-to-region concentration differences for PM are much greater than for ozone and (2) the normal background concentrations of ozone are much higher than for PM.[24]

PM in the atmosphere exists in the form of an aerosol—that is, a suspension of liquids and solids in a gas. At present, particle size forms the basis for the classification of PM. Size is a critical risk determinant and forms the basis for regulatory activity. Somewhat paradoxically, the smaller the particle, the greater the threat to health. Large particles are trapped in the upper airway by nasal hairs and the mucous membranes of the nose, throat, and the upper airway. The smallest particles travel deeply into the lungs and into the *alveoli*, the small air sacs from which oxygen is absorbed by red blood cells and $CO_2$ is eliminated. This is also where PM triggers inflammatory and immunological responses. These responses damage tissues and organs, as shown in figure 7.1. PM is further classified by the mechanism of formation. Primary PM is generated de novo, often by combustion in coal-fired power plants, internal combustion engines, or fires. Secondary PM is formed by physical and chemical reactions among constituents of the atmospheric aerosol. The criteria pollutants—such as oxides of sulfur and nitrogen, formed by burning fossil fuels—are important precursors to these secondary particulates.

PM size would be a simple concept if all particles were spherical and had the same density. Although some particles are spherical and easily described in terms of their size, others are elongated fibers, flakes, or almost any other shape imaginable. Because of this variability, atmospheric scientists use the term *aerodynamic diameter* to describe PM size. Thus, regardless of their size, shape, composition, or density, all particles with the same aerodynamic diameter behave similarly in the atmosphere. More precisely, under the influence of gravity, all particles with the same aerodynamic diameter reach the same final speed as they settle under laboratory conditions. As indicated previously, the smallest particles have the greatest impact on health. These are referred to as $PM_{2.5}$ and have an aerodynamic diameter of 2.5 microns (millionths of a meter) or less. Like ozone, $PM_{2.5}$ concentrations are monitored closely and used to compute air quality indices. The EPA website Enviroflash displays the index for major cities.

A 2009 review article suggests that the relationship between temperatures of the future and PM concentrations is much weaker than the relationship between temperature and ozone concentrations.[25] This conclusion is derived in part from an analysis of data collected between 1990 and 2005 in five cities in the southwestern part of the United States.[26] As expected, since temperature is a force that drives reactions that form ozone, the authors of the aforementioned study found a link between temperature and ozone concentration. There was no link between temperature and PM concentrations. Although there were large variations in the PM concentration, these were thought to be due to between-city differences in the transportation sector, construction, and

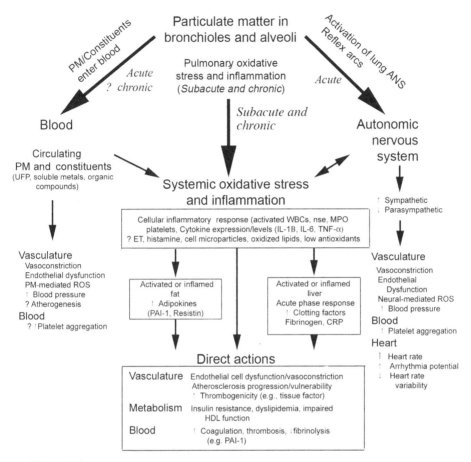

Figure 7.1

Pathways linking cardiovascular disease to particulate exposure. Particulates cause acute, subacute, or chronic effects after they enter the blood; produce systemic oxidative stress and inflammation; and affect the autonomic nervous system. Abbreviations: MPO, meyloperoxidase; PAI-1, plasminogen activation inhibitor 1; WBC, white blood cell; ROS, reactive oxygen species; CRP, C-reactive protein; IL-1B, interleukin 1-beta; IL-6, interleukin-6; TNF-α, tumor necrosis factor alpha; ET, endothelin; HDL, high density lipoprotein; all are molecules mediating or responding to oxidative stress and/or inflammation. *Source:* A. H. Lockwood, *The Silent Epidemic: Coal and the Hidden Threat to Health* (Cambridge, MA: MIT Press, 2012).

industrialization after the passage of the North American Free Trade Agreement (NAFTA), not due to differences in climate.

PM concentrations in the southwest were correlated with relative humidity and may rise and fall in parallel with the water content of the atmosphere.[27] This effect is related to the mechanisms that remove PM from the atmosphere. PM acts as a nidus for the formation of water droplets, which become drops of rain that fall to the ground.[28] As a corollary to this observation, the rate of PM removal from the atmosphere is more highly correlated with the frequency with which it rains than with the intensity of rainfall.[29] Frequent brief rains remove more PM from the atmosphere than rarer more intense rains. Based on predictions of future increases in precipitation in the northeastern United States and decreases in the southwest, PM clearances and hence concentrations are likely to decrease and increase, respectively, in these regions.[30] These results need to be correlated with future emissions of PM in order to gain an accurate perspective on future PM concentrations.

As the result of climate change, severe droughts will probably continue in the southwestern part of the country. As a result, wildfires are likely to become an increasingly serious problem. These wildfires have become a major source of PM discharges into the atmosphere. A glimpse into a future characterized by more wildfires occurred during the 2003 European drought and heatwave. The fires and the resulting discharges of PM into the atmosphere were due to a combination of below-normal rainfall, extreme temperatures, and stagnation of the air that precluded dilution of PM by mixing it uniformly with other parts of the troposphere.[31] A 2008 report of the wildfires in California in 2003 demonstrated the health toll that fire-induced PM can exact.[32] Using satellite-derived PM data, the investigators found increases in $PM_{2.5}$ concentrations of up to seventy micrograms per cubic meter during fires compared to prefire concentrations. These peaks were associated with a 34 percent increase in hospital admissions. Those between the ages of sixty-five and ninety-nine years were the most likely to be affected. During the month of October 2003, the investigators' analysis identified over forty-thousand hospital admissions due to increased $PM_{2.5}$ concentrations.

## Health Effects of PM

Because most PM enters the body via the lungs, that is where many of its effects are found. However, many other organs are also affected, as shown in figure 7.1. In the nose, some very small particles may enter the brain directly via the nerve cells that mediate the sense of smell. These neurons originate in the nose, pass through tiny holes in the bone at the base of the frontal lobes of the brain known as the *cribiform plate*, and then pass directly to the limbic system of the brain.[33] Among other things, the limbic system plays an important role in memory and emotion. In the brain, these particles trigger inflammatory reactions and increase the activity of enzymes that are a part of the immunological system. This causes changes in the brain that are similar to those found in patients with Alzheimer's disease.[34]

PM-induced inflammatory and immunological responses that are similar to those in the brain are also evident in the lungs. These responses produce a condition known as *oxidative stress*, which damages tissues to varying degrees, depending on its intensity.[35] Oxidative stress in the nervous system affects neural reflexes that control blood pressure and other functions that are not a part of conscious activity. Thus, the seemingly innocuous entry of particles into the lungs can result in a constriction of blood vessels and increases in blood pressure. Altered reflex activity also affects heart rate and rhythm and produces atherosclerosis in the arteries to the heart and brain. Atherosclerosis and high blood pressure are major risk factors for strokes and heart attacks and major contributors to the global burden of disease, as discussed in chapter 1. In addition, other changes in the autonomic nervous system help control how our bodies metabolize glucose and insulin, as well as a variety of other abnormalities depicted in figure 7.1.

When the vast epidemiological data linking air pollution and human disease are tallied, it is clear that the four most common causes of death among Americans—heart disease, cancer, diseases of the respiratory system, and stroke—all are caused in part by air pollution. Furthermore, as epidemiological research progresses, the spectrum of disease caused by pollution is likely to widen to include type 2 diabetes mellitus, Alzheimer's disease, and other degenerative diseases of the nervous system.[36]

The Trajectory toward the Future

Climate change seems virtually certain to have some adverse effects on air quality due to expected changes in temperature, humidity, the emission of naturally-occurring VOCs such as isoprene, the flow of air currents, droughts, and related wildfires, and so on. Predicting all of these variables is difficult, so precise estimates are associated with uncertainties. This is particularly true for ground-level ozone. There is much to learn about this aspect of climate change. The literature linking air pollutants with diseases has always been limited by methodology. As techniques for measuring PM have improved, the data about these particles have become more sophisticated and informative. The ability to subdivide particles by size is a perfect example. Three decades ago, many reports referred only to particles with an aerodynamic diameter of ten microns or less. As technology has advanced, it has become possible to measure smaller particles accurately and more frequently. Now, hourly pollution reports focus on those with an aerodynamic diameter of 2.5 microns or less, with the awareness that the smallest particles in that range are the most threatening.

As atmospheric science and epidemiology evolve, it seems likely that analyses of the chemical composition of $PM_{2.5}$ will become more widely available and incorporated into ambient air quality standards. One study of the chemical composition of PM linked higher disease risks to particles produced by burning coal, compared to particles from other sources.[37] The investigators did this by using a statistical technique called *factor analysis* and identified a "silicon factor" linked to particles arising from the earth's crust, a "lead factor" linked to exhaust from motor vehicles, and a "selenium factor" linked to coal combustion. Using additional statistical techniques, the PM from mobile and coal sources was found to increase the risk of death. As a preview of the future, short-duration (thirty-minute) sequential measurements of $PM_{2.5}$ in Killarney, Ireland, demonstrated evening peak concentrations that were more than fifteen times greater than the WHO guideline.[38] These peaks correlated with when workers return to their homes and light coal fires for heat and cooking. No linkages to health outcomes were attempted, although such consequences are likely to happen, as shown by reports linking transient peaks in pollutants to cardiovascular diseases and stroke becoming more widespread, as discussed below. These peaks may disappear or at least become less prominent as the Irish ban on "smoky coal" (bituminous coal) begins to take effect.

Policy always lags behind science—and this is particularly true when air quality standards are revised. Currently, the concentration of ozone is averaged over an eight-hour period. Today's state-of-the-art epidemiological methods have shown, for example, that the risk of an acute stroke rises significantly after a brief spike in the concentration of particulates.[39] There also are other examples in which disease appears after a pollutant spike.[40] An increasing number of Americans have implanted cardiac pacemaker defibrillators, and when these devices detect a potentially fatal heart rhythm disturbance (such as ventricular fibrillation), they attempt to pace the heart to restore a viable rhythm. If that fails, the device delivers a shock to the heart to defibrillate the cardiac muscle. Post-defibrillation data show a close correlation with prior spikes in nitrogen dioxide levels in the air. Among patients who have had ten or more defibrillator discharges, nitrogen dioxide, carbon monoxide, black carbon, and fine particle mass all were linked to the cardiac events. Very large short-duration peaks in the concentration of pollutants may lurk and remain undetected when only the eight-hour average concentration is considered. It is likely that future studies will emerge that firmly establish links between transient peaks in the concentration of air pollutants and diseases. This in turn will require further changes in air quality standards.

# 8

## Violence, Conflict, and Societal Disruption

We can't build a peaceful world on empty stomachs and human misery.
—Norman Borlaug, father of the Green Revolution

The universality of terms such as "hotheaded," "hot-tempered," "hot under the collar," and "hot-blooded" demonstrate the widespread perception that there is a link between temperature and violence. Cultural icons such as Rod Steiger in Sidney Poitier's film *In the Heat of the Night* and the Tennessee Williams play *Cat on a Hot Tin Roof* provide additional support for the validity of the link that seems almost intuitive between heat and violence. The data that examine linkages between heat and violence reinforce that concept. Will the increase in temperatures predicted to occur as a result of climate change lead to more violence? Or will recent trends that show reductions in violence prevail? For example, the FBI reports that there was a 14.5 percent reduction in violent crimes between 2004 and 2013, during which time a total of 368 murders, non-negligent manslaughters, rapes, robberies, and aggravated assaults were reported for each one hundred thousand people in the country.

For the title of his book *The Better Angels of Our Nature: Why Violence Has Declined*, Steven Pinker drew from the last words of Abraham Lincoln's first inaugural speech. The title frames his thesis that there is less violence now than in the past. Lincoln used these words to create an optimistic vision of the future for a nation facing civil war after the secession of the southern states; Pinker uses the same words to help convince us that changes in culture and society have made violence less prevalent.

There is still a great deal of violence among individuals and societies, a fact that Pinker would not deny. It is a rare local television news broadcast that fails to present lurid details of a recent murder. Fear of kidnapping keeps parents from permitting their children to walk to school. Every bank seems to have been robbed. Crimes of all sorts seem to abound. Terrorists recruit suicide bombers, including young women and girls, to strike fear into their target populations. Governments repress, vilify, and may even attempt to exterminate specific populations that differ in characteristics that seem to have little meaning. Although Pinker therefore was not without his critics, he researched and compiled an enormous amount of data to make and reinforce his argument.

Numerous factors lead individuals and societies toward violent actions. A body of evidence that is largely out of sight but growing in acceptance suggests that violence, weather, and climate are closely related. In perhaps the best known of these, investigators examined examples of conflict in which it was possible to infer a causal link between weather or climate and conflict.[1] The authors found that increases in temperature systematically increased the risk of violence. The question is whether evolving cultures and societies of the future will be adequate to promote the trend that Pinker describes and that is reported by the FBI, or whether changes in climate will overwhelm and reverse this trend.

### Individual Acts of Violence

Baseball is a sport characterized by an almost endless list of statistics, including the use of the beanball. A beanball is a pitch thrown with the intent to strike a batter on the bean (where *bean* is a slang term for *head*). The relationship between the beanball and temperature was evaluated in 1991 in a report based on an analysis of 826 randomly selected Major League Baseball games.[2] At that time, the authors found a positive and significant relationship between temperature and the number of batters hit per game. As is often the case, these authors produced a much more comprehensive and detailed follow-up report in 2011, in which they analyzed data from an astounding 57,293 Major League Baseball games to determine whether there was any relationship between the temperature at game time and the probability that a batter would be beaned by the opposing pitcher after a batter on his team was hit by a pitch (deliberately or not).[3] When this happens, baseball tradition and custom have it that "eye for an eye" form of justice must prevail. In this follow-up study, the researchers found that there was indeed a significant interaction between the temperature and what may have been a retaliatory beanball. When the temperature was 57°F and a player on the pitcher's team was beaned in the first inning, the probability that the pitcher would hit a member of the opposing team was approximately 0.22. Under identical circumstances but at a temperature of 95°F, the probability rose by around 23 percent. The risk of getting hit by a pitch also rose steadily when one, two, or three batters had been hit by the opposing pitcher.

This is not the only example of a link between temperature, sports, and violence. Football (American football, not soccer) is a sport that is certifiably more violent than baseball. In an analysis of weather reports, data extracted from over 750 police agencies in the National Incident-Based Reporting System and information gathered from twelve years of National Football League Sunday games, an upset (or an unexpected loss by the home team) was associated with a 10 percent increase in the number of police reports of male-on-female "intimate partner" violence.[4] Heat appeared to be a contributing factor. They defined an unexpected loss as a loss by the home team when it was expected to win by four or more points, according to the pregame point-spread prediction. Wins and expected losses were not associated with additional reports of violence. Additional analyses of the data showed that intimate partner violence was 8 percent higher when the temperature exceeded 80°F.

To illustrate the complexity of the issue, researchers also found that holidays exerted large additional effects on intimate partner violence. Although alcohol and drug abuse are widely accepted as contributing factors in interpersonal violence, there was no conclusive evidence that this played a role in post-football-upset violence. Statisticians frequently create mathematical models of the phenomena that interest them; this group was no exception, and mathematical formulas filled with Greek letters are found throughout their report.

Who hasn't been trapped in traffic? The impact of heat on aggressive behavior by motorists was studied by a group of investigators in Phoenix, Arizona, where temperatures frequently exceed 110°F in the spring, summer, and fall. In this fertile ground for heat-related research, a team deliberately blocked traffic for up to twelve seconds and observed the behavior of the delayed motorists.[5] These intrepid researchers found a linear increase in horn honking as the temperature increased. This effect was accentuated when just those cars with the window rolled down were studied. They hypothesized that these drivers lacked air conditioning and were presumed to be hotter than their fellow motorists who had rolled-up windows. Luckily, that was the extent of the violence they reported. There are places where one might literally risk his or her life by conducting this experiment—or worse, precipitate an episode of road rage directed at the motorists up ahead.

A final and classical example of the link between violence and temperature at an individual level comes from data collected during firearms training exercises for police officers in the Netherlands.[6] This study involved thirty-eight police officers with an average of five years of on-the-job experience. Each officer carried a laser pistol in his or her holster. During the exercises, participating officers were confronted by life-sized television images that showed a suspect holding a crowbar in a threatening position. The video was constructed so that the officer could take evasive action to avoid an attack if he or she chose to do so. The officer's responses were videotaped and scored after the exercise had ended. The temperature on the training site, the independent variable in experiment, was under the control of the investigators. It was set randomly to either 21°C or 27°C. Trainees completed questionnaires that described their reactions to the videos after the exercise. Four outcomes were significantly more probable at the higher temperature: attribution of a negative affect to the subject, the impression that the subject was aggressive, the impression that the subject was threatening, and the tendency to shoot. The officers kept their weapons in their holsters more often in the cooler than in the hotter condition, 15 percent versus 41 percent. Although the officers "shot" the suspect more often in the warmer condition, 62 percent versus 45 percent of the time, this difference did not reach statistical significance.

The investigators made the point that, according to Dutch law, shooting the suspect was not an appropriate action under the circumstances, since the scenario was constructed in a manner that allowed the officer to avoid harm. The report did not describe any ethnic differences among the trainees or the subjects portrayed in the videos, but these are subjects for research with a different but important focus. Based on the number of high-profile shootings by US police officers, it seems quite possible that in a real-life situation, ethnic stereotyping might well magnify any heat-related effect and increase the probability of a perceived threat.

Investigations of the relationships between temperature on one hand and aggressive behavior and violence on the other are often based on so-called field studies—that is, an analysis of actual behavior under real-life conditions. This places constraints on the ability of investigators to control other relevant variables, to quantify other relevant outcomes, or both. Studies performed under laboratory conditions seek to minimize these confounding problems. This is one of the great reasons to conduct laboratory versus field studies.

One study conducted in a laboratory environment involved video games and exercise.[7] The game was similar to Pac-Man, a game in which an operator-controlled token "eats" an object. The level of frustration while playing the game was altered by having some participants use the joystick in the normal position (low frustration) while others played with an inverted joystick (high frustration). Temperatures were set at one of three levels at constant humidity, creating a two-by-three matrix (two frustration levels by three temperatures). In a second, virtually identical experiment, participants stepped up and down on a stool for a minute as a part of the study. The participants completed questionnaires, and physiological indicators (blood pressure and heart rate) were measured at three different times (before exercise and at two points after completion of the exercise). The results of both of these experiments produced increases in what the investigators termed *hostile affect, hostile cognition,* and *physiological arousal* in hot temperatures compared to temperatures that were cooler. They concluded that heat increases the likelihood of bias in a hostile direction during social events that are not typically viewed as hostile or nonhostile.

Craig Anderson, director of the Center for the Study of Violence at Iowa State University, has been one of the most influential and prolific behavioral scientists investigating the relationships between heat and violent behavior. In a 2000 review, his group presented the results of an analysis of aggressive behavior in seven different locations in North America.[8] The group members divided their data into segments that were each one month long. The annual low point for violence occurred in January, when just over 6.5 percent of all assaults took place. The high point came in August, when almost 10 percent of the assaults they analyzed took place. They performed complex analyses of violent crimes in the United States that occurred in 1980. In addition to the crimes themselves and temperature data, they created a factor they called *southernness*. This factor was in part a reflection of what they referred to as a *culture of honor* but was based on more easily quantified elements, such as migration patterns from the South to the North, US Census Bureau data, and the fraction of the 1968 presidential vote that went to George Wallace, an outspoken racist. They also included a measure of social and economic status (SES). Among the three variables, temperature, southernness, and SES, temperature was the one that had the most significant relationship to crime. This was followed by low SES. Southernness did not have a significant relationship to crime, and the researchers concluded that it should be dismissed from consideration. Restated, the significant relationship between temperature and violent crime was stronger than low SES, which was also significant.

They also evaluated data on rape and domestic violence from Minneapolis, Minnesota. The relative rate for both of these forms of interpersonal violence rose smoothly from their lowest points when the temperature was just below 0°F in January to their highest values when the temperature was 95°F in August.

In a 2001 review of the heat hypothesis, which seeks to answer the question "Does excessive heat increase violence?," Anderson makes the following statement: "My colleagues and I believe that most heat-induced increases in aggression, including the most violent behaviors, result from distortion of the social interaction process in a hostile direction."[9] They proposed a four-stage model of behavior to explain aggression and how it is affected by heat.[10] The first stage consists of personal variables, such as aggressive personality, and situational variables, such as uncomfortable heat. These elements feed into the second stage of their model, which they refer to as the *present internal state*. This includes cognitive elements, affective states (such as hostility), and arousal, as measured by heart rate and other physiological variables. These three elements trigger the third stage, which Anderson refers to as *appraisal processes*, or automatic and controlled actions. These appear to be similar to what Kahneman refers to in his book *Thinking Fast and Slow* as system 1, which is fast, emotional, and intuitive, and system 2, which is more deliberative, slower, and more logical.[11] The final stage of the Anderson model is the *outcome*. This is where behaviors may trigger a response, such as a violent act. In this comprehensive review and presentation of original work, the group marshals evidence from multiple sources to support the hypothesis that heat breeds violence.

In their work, Anderson and colleagues address climate change directly.[12] They refer to studies that showed a significant relationship between the hotness of the year and the murder and/or assault rate. When they quantified these relationships and subjected them to a statistical analysis, they found that there were 4.58 additional murder/assault crimes per one hundred thousand people in the United States for an increase in the temperature of one degree Fahrenheit. Using that relationship between temperature and violent crimes and a population of 270 million, a temperature increase of two degrees Fahrenheit led the researchers to predict that there will be twenty-four thousand additional murders or assaults per year in that hotter future. An eight degree increase would produce about one hundred thousand additional crimes.

Anderson injects a note of caution by reminding us that this prediction assumes that social and other systems, such as those that supply food and water, adequate policing, a fair and just government, and other elements that we take for granted in modern society, will remain intact and that other elements of a future climate, such as drought, severe storms, or so on, that place stress on governments and social systems are not present.

## Intergroup Violence

A great deal of the research on violence in societies has been made possible by a database maintained by two Swedish groups: the Peace Research Institute Oslo (PRIO) and the Uppsala Conflict Data Program (UCDP). These data are broken down into several categories: events not intended to damage property or cause injury are classified as *nonviolent events*, and *violent events* implies that the event in question was designed to cause injury or property damage. In addition, a threshold of twenty-five or more deaths is commonly applied to define a *significant event*. Significant violent events include riots—such as antigovernment riots—or repression of populations by a government. Violent events are subdivided into *government-targeted* and *non-government-targeted* events, depending on whether a central government was the objective of the violence. Some food riots fall into the government-targeted category and others into the non-government-targeted category, as discussed in chapter 5. Finally, the two Swedish groups add all violent events together to yield a total for violent events. The UCDP maintains a website that provides interactive maps that allow an interested observer to view the distribution of violence on a country-by-country basis. A quick look at these maps shows that in the 1975–2013 time interval, virtually every nation was involved in violence that was included in the UCDP data set. These data have been used in conjunction with weather and climate data to examine the relationships between weather, temperature, and violence.

Virtually every climate change study predicts that in the future there will be large changes in precipitation in many parts of the world, after which enormous consequences could follow. Most stable societies are adapted to the amount of rain that falls in typical years. Increased amounts of rain can cause flooding that delays planting or harvests, damages crops in the fields, affects livestock, or causes damage to property that is often associated with morbidity or mortality. Droughts also have the potential to disrupt societies in ways that affect the behavior of individuals and groups by causing loss of income, food shortages, price increases, and so on. This relationship has been referred to in the literature as *environmental security*.[13] Investigators have identified at least five mechanisms whereby unexpected departures from the average amount of rainfall may give rise to social conflict:

- Consumers of water may be affected directly. These include individuals or industries that unexpectedly must compete for a scarce resource or face threats posed by floods.
- Price disputes between agricultural interests and consumers may follow droughts or floods. Either droughts or floods are likely to increase the cost of food.
- When areas become uninhabitable, residents must move, usually into a city. Competition for jobs, housing, and services, such as those provided by law enforcement and utilities, may become sources of tension that can trigger conflict. Some refugees are likely to cross national boundaries.
- Confronted by internal problems related to the supply of commodities, governments may intervene in markets in one or more ways that create tension. Interventions are often related to exploitation of

markets to increase profits or political advantages. This is more likely to occur in countries run by corrupt officials or by a single strongman.

• Macroeconomic effects in a nation or region may take a human or financial toll.

Sadly, Africa has proven to be a laboratory for investigating violence and its relationship to a variety of stressors, including climate. Using the UCDP and PRIO data, investigators have conducted a detailed analysis of over six thousand social conflicts in Africa that took place over a twenty-year period.[14] These investigators combined these data with rainfall data for each country from the Global Precipitation Climatology Project. Their analysis showed that there was a significant positive association between all forms of political conflict with deviations from the average annual rainfall. It did not matter whether there was too much or too little rain. However, civil wars and other violent conflicts were more common during the wet extremes than during the dry extremes.

Because rainfall in Africa ranges wildly between Egypt, where 3.1 cm of water falls in a typical year, to Sierra Leone, where the annual expectation is 233.3 cm, the investigators normalized rainfall for each country. This process results in a similar standard value for each country, as well as similar values for deviations from this average. They use the term *rainfall deviation* to describe these normalized data. This procedure makes it possible to include many nations with widely divergent average rainfall values in a single analysis. After the normalization was complete, they combined data from multiple countries and expressed the percentage by which conflicts changed as a function of departures from this normalized amount of rain. The results of this quite complex analysis are shown in figure 8.1. The strongest association was found for the total number of events, as defined previously, and rainfall deviations. For example, an increase of just over 33 percent (one standard deviation) in the normalized rainfall was associated with a 6.1 percent increase in social conflicts, whereas the decrease of a similar amount was associated with a 9.2 percent increase. More extreme deviations in rainfall were, by definition, less common, but were associated with larger increases in the number of conflicts. When rainfall increased by just over 47 percent and almost 50 percent, conflicts increased by 38.1 percent and 103.5 percent, respectively.

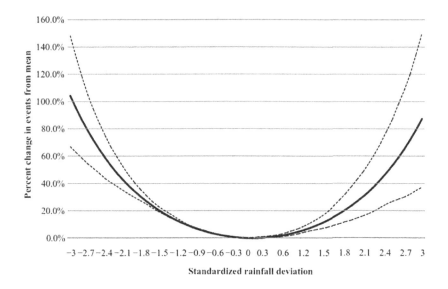

Figure 8.1

Marginal effects of rainfall deviations on total social conflict events. Increases and decreases in the expected amount of precipitation are associated with increases in violence. The solid line is the estimated percentage change; the dashed lines represent the 95 percent confidence interval. Reproduced with the permission from C. S. Hendrix and I. Salehyan, "Climate Change, Rainfall, and Social Conflict in Africa," *Journal of Peace Research* 49, no. 1 (2012): 35–50.

Decreases of similar amounts were also associated with increases of 30.7 percent and 88.2 percent, respectively. In other words, they found a U-shaped curve: A departure from expected amounts of rainfall, whether it was an increase or decrease, was associated with an increase in the probability of violence.

Some of these fluctuations in rainfall and attendant violence are related to predictable fluctuations and prolonged warming of the sea surface along the western coast of South America, known as El Niño, and the atmospheric pressure component associated with the temperature changes, known as the Southern Oscillation.[15] Collectively, these two phenomena are referred to as ENSO (see also chapter 4 for ENSO effects on malaria). Variations in ENSO are associated with climate fluctuations in many tropical regions worldwide and along the northwestern coast of North America even though they are at some great distance from the Pacific Ocean where ENSO arises. A relationship between distant phenomena is referred to as a *teleconnection*, a term used by atmospheric scientists to describe features of the climate that are related to each other over large distances, often thousands of miles.

Researchers have correlated fluctuations in ENSO with the probability of a civil conflict in these regions that are teleconnected to ENSO.[16] They found that the probability for a civil conflict doubles during intervals when the influence of El Niño is strong compared to intervals when it is weak (referred to as La Niña). They conclude that from a statistical perspective, changes in the ENSO pattern were linked to 21 percent of conflicts that developed between 1950 and 2004. Although additional studies should be performed that test the hypothesis that ENSO and conflict are linked, this study at least raises the possibility that the researchers' conclusion is accurate. The potential importance of seemingly disparate linkages was addressed in a discussion of vulnerable populations in a landmark paper published in 2015 in *The Lancet*, in which the authors emphasized the "interconnected nature of climate systems, ecosystems, and global society."[17]

Increases or decreases in rainfall commonly lead to reductions in the harvest of agricultural commodities and increases in the cost of food. As a corollary to the study of violent and nonviolent social conflict, another report focused on food riots and food prices in Africa and the Middle East.[18] The results of this study are shown in figure 8.2, which shows clustering of food riots around peaks in the Food Price Index as reported by the Food and Agriculture Organization.

The World Bank Group monitors global agricultural production and prices, reporting these results periodically in its newsletter, *Food Price Watch*. The May 2014 issue addresses the nexus between food supplies and violence by addressing food riots. They note that others have used the term *food riot* imprecisely and with no real consensus as to its definition. Is loss of life necessary? How long must a disturbance last? What about peaceful demonstrations? The World Bank begins with a proposed set of defining criteria: "A violent, collective unrest leading to a loss of control, bodily harm, or damage to property, essentially motivated by a lack of food availability, accessibility or affordability, as reported by the international and local media, and which may include other underlying causes of discontent."[19] The organization would exclude most of the 2011 Arab

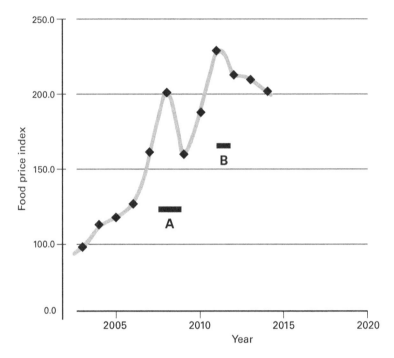

**Figure 8.2**

Food riots and the Food Price Index. The Food Price Index was obtained from the Food and Agriculture Organization (http://www.fao.org, accessed January 14, 2015). In time interval A, food riots in Somalia, India, Mauritania, Mozambique, Yemen, Cameroon, Sudan, Ivory Coast, Haiti, Egypt, Somalia, and Tunisia caused approximately eighty-nine deaths. Many more died in time interval B, during riots in Mozambique, Tunisia, Libya, Egypt, Mauritania, Sudan, Yemen, Algeria, Saudi Arabia, Oman, Morocco, Iraq, Bahrain, Syria, and Uganda. These riots claimed more than twenty thousand lives, with more than half of those lost in Libya. Data from M. Lagi, K. Z. Bertrand, and Y. Bar-Yam, "The Food Crises and Political Instability in North Africa and the Middle East," 2011, http://papers.ssrn.com/sol3/papers.cfm?abstract_id=1910031.

Spring uprisings, because food was not a major aspect of these events, but it would include prerevolution events in Algeria and Tunisia, because inflated food prices were primary precipitating factors. The World Bank subdivides food riots into two types: Type 1 riots are the most frequent and are directed at governmental authorities; Type 2 riots are devoid of political aspects, are not directed at governments, and typically target food repositories such as warehouses, shops, and trucks. The authors of the *Food Price Watch* report note that food prices are frequently not the exclusive cause of violent unrest. Predisposing conditions also are common precursors to riots. These conditions include poverty, weak or inadequate governments, and an absence of planning for disaster. There does not appear to be a simple answer or solution to what may become an increasingly common and difficult problem as population growth, rising temperatures and increases or decreases in rainfall alter the socioeconomic landscape of the future.

## Societal Collapse

The studies summarized in the first part of this chapter addressed the short-term effects of weather and violence, such as horn honking during hot spells in Phoenix, controlled environments during police weapons training, and laboratory environments in which the temperature was manipulated during the execution of tasks. Then, I considered events that last for a longer period, such as the relationship between El Niño and the Southern Oscillation, and violence in Africa and across the globe, food riots, and other forms of civil unrest. When considering the total collapse of a civilization, which may take a substantial period—decades or even centuries—considerations of weather (short term) turn to considerations of climate (long term).

In his book *Collapse: How Societies Choose to Fail or Succeed*, Pulitzer Prize–winning author Jared Diamond develops a thesis to explain the end of past civilizations.[20] He postulates five elements that operate to varying degrees:

- *Environmental damage caused by inhabitants*. This may take the form of stripping an island of its trees, or agricultural practices that cause a loss of topsoil that leads to crop failures.
- *Hostility of neighbors*, often preceded by weakening of the affected society. Here, Diamond cites the fall of Rome to barbarian invaders. Roman society was weakened and eventually fell after a series of unsuccessful invasions by hostile neighbors.
- Related to the hostile neighbor scenario, *a reduction in support from a neighbor*.
- *Inadequate responses to societal problems*. Here, Diamond cites the deforestation of Greenland by the Norse settlers and a similar loss of

trees on Easter Island as examples of how environmental damage combined with inadequate societal responses led to collapse. By contrast, the Japanese and other island civilizations developed programs to buttress their societies, making them stronger.

• *Climate change*, the subject of this book.

These are undoubtedly not independent variables. For example, a prolonged drought may weaken a society, which makes inadequate responses to problems, which eventually leads to a takeover by another. Climate change may combine with self-inflicted environmental damage, which weakens a society, which makes poor decisions concerning resource allocation. These decisions, in turn, lead to conditions under which continued success of the society is impossible. Based on what we know about the relationship between stress and violence, it seems likely that interpersonal, intrasocietal, and intersocietal violence increase as a society collapses.

This appears to be the rationale exercised by Solomon Hsiang and his colleagues in their quantitative analysis of climate and conflict that spanned twelve thousand years, beginning ten thousand years BCE and continuing to the present.[21] They used a broad definition of conflict that ranged from examples of short-term violence between individuals, such as the police-training episode reviewed earlier, to political instability and civil war. Their comprehensive study drew on data collected by multiple disciplines in the social sciences, including psychology, criminology, economics, archeology, geography, history, and political science. They evaluated sixty studies of forty-five sets of data about conflicts published in twenty-six journals by over 190 researchers. These scholars had well-defined inclusion and exclusion criteria in their study. Only those publications and data sets that made it possible to make quantitative assessments were used. The researchers ultimately concluded that climate had a major influence on conflict between individuals and societies as a whole. When they focused on studies published after 1950, they found that "climate's influence on modern conflict [was] substantial and highly significant." A temperature increase of about 34 percent from the mean of the time (one standard deviation) was associated with a 4 percent increase in interpersonal violence and a 14 percent increase in intergroup conflict.

To illustrate the importance of the work done by multidisciplinary researchers in the examination of the fate of early American civilizations,

I turn to expeditions sponsored by the National Geographic Society, as reported in 1929 in their journal, *National Geographic*.[22] Andrew Douglass, the leader of the National Geographic expeditions, began his career as an astronomer who specialized in sunspots. Based on earlier reports, he thought that his astronomical expertise would allow him to link the study of growth rings to variations in sunlight caused by fluctuations in the number of sunspots. This hypothesis was ultimately shown to be untrue, but his observations of tree rings were of seminal importance.

Douglass is now revered as the father of a scientific discipline known as *dendrochronology*. His efforts have made it possible to use tree ring data to establish accurate links between rainfall and seminal events in various civilizations, such as those that once flourished in the Southwestern United States, the Angkor area of Southeast Asia in what is now Cambodia, and elsewhere. He recognized that trees are not "mechanical *robots*,"[23] but are living things, and a tree's "food supply and adventures through life all enter into its diary." He wrote, "Trees are nature's rain gauges."[24] He describes finding a log near Show Low, Arizona, with the field designation HH-39 that, in his words, proved to be the Rosetta stone for tree-ring research. (*Note:* The name Show Low is believed to be the result of a lengthy card game in which one player bet his 100,000-acre ranch on the turn of a card. His opponent showed the lowest card and won.) The HH-39 log unlocked the secrets of tree rings and rainfall. Using tree ring spacing that can be accurately linked to rainfall, it can be shown that widely spaced rings indicate large amounts of rain and narrowly spaced rings indicate drought. This turns out to be a much more precise dating method than carbon-14 ($^{14}$C) dating and other commonly used techniques. Using the science of dendrochronology, it is possible to create timelines that link salient historical information about a society with the exact year of occurrence. Douglass's examples linked specific events in the Southwest to the time of the Sixth Crusade, the reign of Charles Martel, and other events that took place in other parts of the world. Later in this chapter, I will discuss the use of tree ring data to chronicle the decline of the Angkor civilization.

Diamond also looks at the American Southwest. He tells the story of the Anasazi, who came to live in what is now called the Chaco Canyon, located in the northwestern portion of New Mexico.[25] This society had its heyday between 600 and 1200 CE. From a series of seemingly disparate observations that are based on studies of packrat feces, isotopes of strontium in logs, and other trace evidence of the society, scientists found that the Chaco Canyon site was the focal point of a society that spread outwards. As the community grew, local resources were exhausted. More trees were cut than could be replaced by normal growth. In this time of plenty, the region became unable to produce enough food locally. The inhabitants therefore had to import food from adjacent areas where water and the soil were able to support the growth of corn, their principal crop. Trees were cut in distant forests and used as building material. The beginning of the end came with the onset of a severe drought that led to the collapse of agriculture. Starvation conditions ensued, and people began to eat mice rather than deer, which had vanished from the area. Within-group strife and warfare, including evidence of cannibalism, marked the final days of this once-prosperous society.

Other American civilizations have been studied using a variety of other archeological tools. With a combination of validated strategies, anthropologists and geologists have studied human populations in what is now known as the Bighorn Basin in Wyoming.[26] These scientists combined $^{14}$C radiocarbon dating, an analysis of fossilized pollen samples (used to form another critical database for the study of ancient civilizations), and the analysis of stable oxygen isotopes in corals (similar to the oxygen isotopic studies discussed in chapter 2) to study the remnants of the societies in that area. They found evidence for five episodes in which there were major increases or decreases in the population of the area over a time span of thirteen thousand years. Using artifacts left behind by these inhabitants and temperature and rainfall data, they were able to link the rise and fall of this population to climate. They reported that 45 percent of the variation in the number of people living in the area could be explained by reconstructed concordant temperature and moisture records. They found that on average there was a three-hundred-year interval between a climatological change and a change in the size of the population. Other studies show that this is the predicted length of time between a major change in the climate and an effect on the population of the area. Whatever the explanation, these investigators report clear evidence that the population adjusted to environmental conditions on a continuing basis.

On the other side of the world, in Angkor—the capitol city of the Khmer Empire in what is now Cambodia—another group of scientists used climatological data to time the rise and fall of this fourteenth- and fifteenth-century civilization.[27] Their study relied heavily on tree ring data. From 7.5 centuries of evidence, they were able to establish the dates for a series of droughts and monsoons. They found that droughts lasted decades and were interspersed with periods of much higher than usual rainfall during intense monsoons. Droughts damaged agricultural production and had adverse effects on intricate systems of canals and components of the Angkorian water-management infrastructure. Floods from intense monsoons also damaged the water-handling systems by inundating them with sediment. The scientists linked these environmental disasters to changes in the surface temperature of the Pacific, concluding that a warm ocean and El Niño events led to drought on the time scale that coincided with the downfall of the Angkorian civilization.

## The Trajectory toward the Future

The four climate-modeling scenarios used by the IPCC predict different degrees of warming by the end of the century.[28] The amount of warming, of course, depends on the efforts to mitigate climate change—that is, to reduce greenhouse gas emissions, particularly in nations that emit huge amounts of these gases. The representative concentration pathways used to describe the climate of the future (see chapter 2) include considerations of how land is used, whether or how much populations grow, the characteristics of energy use and production, and social and economic factors in addition to greenhouse gas emissions.[29] Based on the current responses (or nonresponses) to the threats posed by climate change, it seems virtually certain that significant worldwide warming will occur and that there will be substantial changes in rainfall across the globe. Some regions will experience large increases in precipitation, whereas others will be gripped by megadroughts. Neither of these precipitation alternatives holds the promise of a better world.

Most researchers who have investigated the relationships between climate and weather on one hand and violence, conflict, and social disruption on the other have looked to the past. It is difficult enough to develop, validate, and apply the varied tools of history, anthropology, economics, geology, sociology, psychology, physiology, and others to the study of what has happened; it is even more difficult to predict what will happen in the future.

Hsiang and his colleagues are exceptions to this rule.[30] Operating from the premise that past behavior is the best predictor of future behavior, they have risked a look into the future. Recall that they found that temperature increases were associated with increases in interpersonal violence and intergroup conflict. Using an entirely plausible model of the future climate, they found that huge portions of the inhabited world could expect mean temperatures to increase significantly. Some of these regions are clustered about the tropics and subtropics, where the El Niño and Southern Oscillation climate–conflict linkages are the strongest.[31] These are also regions where childhood malnutrition is the highest and tropical diseases are rampant. In other words, these are the parts of the world where the five elements that Diamond believes lead to societal collapse are the strongest. These elements may become the factors that determine the fates of these societies.

# 9

## Economic Considerations of Climate Change and Health

The true economic impact of climate change is fraught with "hidden" costs.
—Matthias Ruth, Dana Coelhl, and Daria Karetnikov, in *The US Economic Impacts of Climate Change and the Costs of Inaction*[1]

### General Considerations

The 2015 Nobel Prizes in Medicine were awarded to Satoshi Omura, Youyou Tu, and William C. Campbell for their research that led to effective treatments for parasitic diseases, notably malaria and river blindness also known as onchocerciasis or Robles disease. Malaria continues to cause millions of deaths each year. By contrast, river blindness is on the verge of extinction. The Carter Center established by President Jimmy Carter has led the efforts to eradicate this disease. As of this writing, President Carter suffers from metastatic malignant melanoma and faces an uncertain future. However, always the optimist, he is reported to have said that he hopes to outlive the last parasitic worm (*Onchocerca volvulus*) that causes this form of blindness. Although not noted specifically in the news that attended the announcement of the Nobel Prize, it is virtually certain that the cost of treatments for these diseases is substantially less than the cost of the diseases they treat. Economics plays an increasingly important role in making healthcare decisions. In their report on the economics of climate change in the United States, Ruth and associates lament the fact that short-term costs associated with mitigation and adaptation to climate change, frequently expressed in terms of jobs and dollars, outweigh the larger costs associated with doing nothing.[2]

The authors of the chapter on economic considerations in the IPCC Fifth Assessment Report point out that there is a continuous, graphical relationship between the costs of climate change (damages) and the cost of adaptation (actions taken to minimize damages).[3] At one extreme, additional spending produces smaller and smaller levels of adaptation. Also, technological limits constrain adaptation. No matter how much more is spent, additional adaptation is not possible. Some damages are unavoidable. Near the other end of the curve, relatively small expenditures result in substantial reductions in damages. A middle ground lies between these extremes. There is a balance of sorts between what can be done and the funds available to make changes. The fifth assessment authors refer to this state as *what we want to do*. The position on this part of the curve involves weighing alternatives and making decisions— decisions that are almost always driven by conflicting political perspectives. In the United States, we see this situation all the time as environmentalists and supporters of public health urge policymakers to work as hard as possible to protect human health and the environment. The climate change deniers are at the other end of the spectrum. They claim that climate change is an elaborate hoax and that it is absurd to spend money on a problem that does not exist. They are joined by many industry groups who claim that we cannot possibly afford to make these adaptive changes because of the costs (to profits) and the number of jobs that would be lost.

The task of affixing a price tag to the costs associated with climate change is akin to capturing a will-o'-the-wisp. If an accurate estimate of costs exists at all, like a will-o'-the-wisp, it may be seen best at night flickering over swamps or bogs. In a 2009 report titled "Assessing the Costs of Adaptation to Climate Change," the authors make this point more elegantly by noting that although there may be a "comforting convergence" of data presented by authoritative sources such as the World Bank and the United Nations Development Program, it would be unwise to rely on what seems to be a general agreement among reports. They reach this conclusion because "(i) none of these [reports] are substantive studies, (ii) they are not independent studies but borrow heavily from each other, and (iii) they have not been tested adequately by peer review in the scientific or economics literature."[4] In their critique, the authors note that the analysis by United Nations Framework Convention on Climate Change (UNFCCC) was restricted to just three health scenarios: malaria, diarrheal diseases, and malnutrition data from the 2004 Global Burden of Disease Report. In their view, a much broader assessment was warranted. The important task of creating accurate estimates of the cost–benefit ratio on each point of the curve described previously appears to be an important but wide-open field for data-driven research that can lead to evidence-based decisions.

A more recent report titled "Risky Business" was produced by a group that included Michael Bloomberg, Henry Paulson Jr., and George Shultz. The title echoes that of the 1983 comedy in which Tom Cruise converts his home into a bordello while his parents are away in order to demonstrate success in business and gain entry into Princeton. Presumably, the real-life businessmen who spearheaded the "Risky Business" report wished to make a profit and at the same time draw attention to the risks we as a global society are taking with our planet. This serious white paper is based on the companion, much more detailed report titled "American Climate Prospectus: Economic Risks in the United States."[5] Although it is not peer reviewed in the traditional sense, it was prepared and reviewed by outstanding scientists and economists and presents a broadly based recent economic analysis that is focused on the United States. In general, it ignores worldwide health and other climate-related problems. The report's bottom line is that climate change will impact many segments of the economy and the costs will be high. As a corollary, mitigation is warranted.

## Monetizing Heat Illnesses

It is necessary to make myriad assumptions in any attempt to monetize any illness or death—a task that is not for the faint of heart. Many different outcomes are possible, depending on which assumptions one chooses. In the approach to monetizing heat illnesses, various agencies have produced cost estimates for heat-related illnesses. Here are a few of the results.

A 2012 posting by the California Compensation Fund provides one perspective (http://www.dir.ca.gov/dosh/heatillnessinfo.html, accessed February 18, 2016). California law requires shade and water at outdoor work sites on "hot days," but it is not clear what is meant by hot days.

However, violations found at the time of an inspection can result in fines or shutdowns. After two cases of heat illnesses in July 2011, two farm labor contractors were fined more than $135,000. It seems likely that many additional violations did not result in fines. In that same year, the California Occupational Safety and Health Administration conducted more than three thousand inspections of work sites and issued 919 citations for various work rule violations (http://www.statefundca.com/home/StaticIndex?id=http://content.statefundca.com//news/Feature Articles2012/050312-HeatIllness.asp). These citations cost employers more than $500,000. In other words, the inspectors found a violation one-third of the time, and the fines were likely to be higher than the cost of complying with worker protection regulations.

A Washington State report presented the results of an analysis of 483 workers' compensation claims for heat-related illnesses during the 2000–2009 interval.[6] There were 3.1 claims for every 100,000 full-time equivalent employees. More than three lost days of work occurred in 10.2 percent of the claims. These claims cost the state fund an average of $3,682 per claim, for a total of $1.78 million.

Serious heat-related illnesses that are not immediately fatal require hospitalization. A 2005 survey of US community hospitals uncovered approximately 6,200 hospitalizations for heat-related illnesses.[7] As we might expect, the poor were the most seriously impacted, with hospitalization rates around twice as high as those for the wealthy. The average cost of a hospitalization was $6,200 per stay, for a total cost of just over $38 million. These hospitalization cost estimates, now a decade old, seem low. It is difficult to imagine that contemporary hospitalizations, with increasing use of expensive imaging, multiple consultants, and reliance on intensive care, would be similar or even proportional to inflation-based increases.

An unprecedented heat wave affected much of California during the last half of July 2006. Daily maximum temperature records were set at seven locations, and daily minimum temperatures that were higher than normal were recorded at eleven locations. The authors of this study found evidence for a substantial increase in heat-related morbidity during that period of record-breaking heat.[8] Compared to two reference time intervals, the heat wave led to over sixteen thousand excess visits to emergency rooms and over 1,100 hospitalizations. Records from coroners and medical examiners produced evidence for 140 heat-related deaths. This is virtually certain to be a more accurate number than the one found on the NOAA website, referenced in chapter 3, which lists just sixty-five heat-related deaths in California for the entire year of 2006. Children less than four years of age and adults older than sixty-five were at the greatest risk. Morbidity was related to electrolyte disorders and kidney failure, presumably due in part to dehydration, cardiovascular disease, and diabetes. Around thirty-seven million people lived in the study area, so there were around forty-three emergency room visits per one hundred thousand people. Even if the $6,200 cost per hospitalization is correct, the emergency visits cost just below $100 million.

The disability-adjusted life year (DALY) is a widely used measure of monetary impacts of health-related issues. It extends the cost of a death reported by another statistic, known as the years of life lost (YLL), to include the burden of disease due to disability or poor health. Unfortunately, neither the DALY nor the YLL measure appears to be applied widely to heat-related death and morbidity. In the 2010 Global Burden of Disease study, heat is lumped together with burdens due to fire and hot substances.[9] In this study's report, there were 330,000 deaths in this category among all ages for all nations. This was an increase of 23 percent compared to the earlier, 1990 study.

Although the authors of the reports discussed previously generally do not monetize their results, they provide a starting point for this task. As discussed in chapter 3, the 2003 European heat wave may have claimed as many as seventy thousand lives. In its 2010 report to Congress on the benefits and costs of the Clean Air Act, EPA used $7.4 million as the value of a statistical life (VSL) in 2006 dollars.[10] Applying that number to the European heat wave yields a cost of over $500 billion. This is almost certainly far too high, as it is probable that many of the deaths were going to occur in the near future, a shift that is referred to as a *displacement* or *harvesting effect*. However, it is also certain that many of those who died were infants and children. Those deaths were premature. For this group, the VSL may be too low. Regardless of how one computes the cost, the heat wave exacted a substantial toll on Europeans. The California and Chicago heat waves that claimed 140 and approximately seven hundred deaths respectively would yield costs of just over $1 billion and $5.2 billion respectively in terms of VSL. The good news is that the knowledge gained by the examination of these events shows a path to improved health outcomes and reduced costs. This is apparent in chapter 3, in which we showed that air conditioning and improved contact and service strategies directed at the most vulnerable individuals yield large benefits.

## Vector-Borne Illnesses

### Dengue

*Dengue* is one of the world's first hitchhikers. The author of a history of the disease notes that dengue had been known for a long time before it was distributed throughout the tropics in the eighteenth and nineteenth centuries.[11] Both the mosquito vector *Aedes aegypti* and the dengue virus were stowaways in the water supplies of sailing vessels. When the ships made landfall, the mosquitoes and their viruses jumped ship. Local epidemics followed at relatively infrequent intervals, presumably because of the slowness of the sailing vessels that moved the mosquitoes and the virus. This changed dramatically during World War II, when troop movements in Southeast Asia carried an unwanted passenger along with the supplies needed to maintain the troops. The first recorded modern epidemic of dengue hemorrhagic fever struck Manila in the Philippines in 1953 to 1954. Spread of the disease to Thailand, Singapore, and Vietnam followed.

In a subsequent editorial, the aforementioned author wrote that dengue had been "considered an unimportant public health problem because mortality rates were low and epidemics occurred only infrequently ... [and after World War II] great progress was made in controlling infectious diseases of all kinds, especially vector-borne diseases, and the war on infectious diseases was declared won in the late 1960s." The editorial continued, stating, "In 2012, dengue is the most important vector-borne viral disease of humans and likely more important than malaria globally in terms of morbidity and economic impact."[12]

The authors of a 2012 paper that was the subject of the aforementioned editorial reported on a study of the economic impact of dengue on Puerto Rico.[13] After the number of cases reached a record high, the authors conducted a comprehensive study of one hundred patients who had laboratory confirmation of a dengue virus infection between 2008 and 2010. By extrapolating their results to all of Puerto Rico, the authors concluded that the annual cost of the disease between 2002 and 2010 was $38.7 million, of which 70 percent was associated with those aged fifteen years or more. This cost rose to $46.45 million per year (or $12.47 per capita) when additional costs such as disease surveillance and mosquito control activities were included. Thus, the total for the nine-year period studied was $418 million. Not surprisingly, those ill enough to require hospitalization accounted for 63 percent of the costs of dengue. Those who died consumed 17 percent of the costs. Individual households bore 48 percent of the costs, with government-funded programs picking up 24 percent and insurance another 22 percent. Employers were impacted the least, accounting for only 7 percent of the costs. The study did not account for the value of lost years of life. Had it done so, the numbers would have been even higher.

As large as the Puerto Rican numbers are, they are dwarfed by those in a report on the economics of dengue in India.[14] Before this study, the official governmental data suggested that there was an average of 20,474 cases per year between 2006 and 2012. Based on better data from the Madurai district and the analysis of an expert panel, the actual average should have been around 5.8 million cases, for an underreporting factor of 282! Total direct medical costs were estimated to be $548 million per year in US dollars. One-third of the cases required hospitalization, a number that appears to be too high based on usual disease severity estimates. This detail suggests that there may be even more cases than the government suspected. When nonmedical and indirect costs are added to the medical tally, the total economic burden climbs to around $1.11 billion, or $0.88 per person per year.

In their review of the literature, the principal author of the Puerto Rican economics paper wrote that the World Health Organization (WHO) reported a thirty-fold increase in the number of cases in the past half century.[15] The WHO estimated that there were between one hundred and two hundred million infections per year, most of which were clinically silent; thirty-four million cases of symptomatic dengue fever; and two million cases of the most severe form of the disease, dengue hemorrhagic fever. They concluded that dengue could be the most important vector-borne viral disease in the world.

One of the stated purposes of the Puerto Rican study was to inform policy makers; the Indian study has similar implications. Policy makers need to be informed because a vaccine is on the horizon. The fact that the Puerto Rican study was funded by Sanofi Pasteur, the developer of a vaccine that I discuss in chapter 10, caused only a minor twitch of the editorialist's eyebrow.[16] In spite of the potential for a conflict of interest, he dismissed this concern based on the detail of the work presented, the reputation of the scholars who performed the study, and the fact that the work was done via a contract with a university (Brandeis) rather than the pharmaceutical company itself.

The Puerto Rican study is narrow in its scope but thorough in its analysis and findings. This is something of a rarity in the literature. The need for studies of this nature was pointed out by the authors of a comprehensive review of dengue published in 2011.[17] The review authors identified 748 publications that dealt, in one way or another, with the economics of the disease. They found such disparity among definitions, survey methods, sampling periods (some sampled only during epidemics, others more broadly), what was included in the costs, and myriad other factors that it was difficult to make a meaningful generalization concerning costs to various economies. Among those that monetized costs, they found a Nicaraguan study that estimated total costs per year in that country of $2.7 million, for a mere $44 per case. A Thai study listed cost at $12.6 million per year. Broken down, this amounted to $118 per child in Bangkok and $161 per adult; costs in another city were lower. In another compilation of costs from Brazil, Cambodia, El Salvador, Guatemala, Malaysia, Panama, Thailand, and Venezuela, total costs were tallied at $851 million per year. Broken down, this amounted to $248 per nonhospitalized case, rising to $571 for hospitalized cases. Anyone with even a passing familiarity with the costs of medical care in the United States will realize that these figures hopelessly underestimate what the costs of care for dengue would be in this country. Luckily, there currently are not many cases of dengue in the United States—but this could change quickly and with little warning. The same is true for instances of microcephaly caused by Zika virus infections. The lifetime costs associated with children born with this severe disability are huge. This is particularly true for children who are institutionalized and for those who are cared for at home, caregivers make large economic sacrifices.

The authors of the 2011 review also tallied data about the cost of vaccination.[18] Cost data are one of the main factors that spur economic research. Policymakers are interested in cost–benefit data and want to know the cost of the disease and the cost of vaccinating the population at risk. Cost estimates for vaccinating citizens in countries with highly varied levels (and costs) of healthcare must be obtained. The review authors assumed that there were 1.2 billion people at risk and that a two-dose regimen would cost eighteen dollars per dose. This included the cost of the vaccine and the cost of administration. Therefore, universal vaccination would total $43.2 billion. However, the Sanofi Pasteur vaccine was given in three doses, which would raise the cost to nearly $65 billion. This figure assumes that once a person is vaccinated, he or she is protected for life; this may not be the case. Finally, the Sanofi Pasteur vaccine was only about 65 percent effective in the Latin American study that will be discussed in chapter 10.[19]

## Malaria

Malaria has two faces: it is both a disease of the poor and one that causes poverty. Countries where malaria is common are often so poor that they are unable to take the steps needed to control the disease. Thus, it continues to ravage the affected nation. It is also a tremendous killer, particularly among children, as detailed in chapter 4. By destroying the lives of so many, either by death or the associated morbidity of chronic disease, malaria deprives nations of their futures. This makes it extremely difficult to make economic progress at a rate that is characteristic of comparable nations, resulting in a continuous cycle of poverty breeding poverty. For these and other equally good reasons, the United Nations has included controlling malaria in its Millennium Development Goals, as discussed in chapter 1. As a part of this process, the World Health Organization publishes an annual malaria report in December of each year.

The authors of a review of the economic and social burden of malaria began their article with a reference to Darwinian principles.[20] They referred to *sickle cell anemia*, a genetically-determined disease that is often fatal if the individual has two genes for it (homozygous), but protective against malaria if the individual carries only one gene (heterozygous). In other words, the rewards for carrying just one copy of the gene are so great that the trait persists in spite of the fact that it carries a significant risk when an individual has two copies of the gene. These unfortunate individuals are likely to experience morbidity and mortality associated with sickle cell disease.

For many of the poorest countries, the costs of malaria are the highest, no matter how one examines the problem. Data from a 2004 comprehensive review of the economics of malaria in developing countries brings focus to this issue.[21] Per capita costs per month ranged from $0.46 nationwide in Malawi to $5.98 in urban parts of Cameroon. Costs per patient per episode were $2.09 for all cases in rural Ghana and $3.28 in rural Sri Lanka. When the cases in rural facilities in Ghana were broken down by severity, mildly affected individuals cost $3.72, whereas severe cases cost $7.38. Although these costs may not seem high, when expressed as a percentage of income lost due to loss of labor of patients and their caregivers, the costs soar. In Malawi, nationwide, this so-called indirect cost consumed 2.6 percent of household income each year. In rural Sri Lanka, the annual cost per episode was 6 percent of household income. Citizens of some African nations, such as South Africa, Lesotho, and Mauritius, spend less than 5 percent of their household incomes on malaria-related expenses. However, for twenty-six other countries the fraction hovered between 15 percent and 20 percent.[22] Thus, the financial and associated social burdens of the disease accounted for a major portion of household expenses and was borne by those who could least afford it.

The authors of a 2002 review of the social and the economic costs associated with malaria pointed out the shortcomings of a strategy based on a determination of the cost an individual incurs during an episode of the illness, the number of episodes per year, and the population.[23] The scope of their analysis is much broader and includes changes in household behavior and effects on trade, tourism, foreign investment, and similar economic realities.

Malaria affects families in various ways. When parents expect that one or more of their children may die from malaria, there is a tendency to have more children. This is consistent with the *child survivor hypothesis*, the idea that reproductive decisions are based in part on a desire to raise a certain number of children into adulthood. Some economic resources are expended on children who die, and it is probable that the fixed family income that is spread over a larger number of children reduces the per-child investment in education. This fact is almost certain to have a larger impact on girls than boys. Among the children who survive, many will miss a significant number of days at school due to malarial illness. In Kenya, estimates suggest that malaria causes children to miss 11 percent of all school days. This detail has additional impacts on failure rates, dropouts, and the need to repeat a grade. The per capita GDP is lowered in these larger families, because only the adults produce virtually all money counted in this economic indicator.

Children who survive malaria are more likely to be anemic and undernourished. This in turn impairs development, including intellectual development. Cerebral malaria exacts an additional toll, as children who survive may have learning disabilities or other forms of neurological impairment including epilepsy.

In countries where malaria is endemic, adults develop partial immunity to the disease due to constant reinfection and the immunological response it generates. This immunity may wear off among adults who leave the country for education or other purposes. When they come home, they are susceptible to reinfections and are prone to develop severe manifestations of the disease due to the partial loss of immunity.

Travel, tourism, and foreign investments are lowered by the risk of infection in countries where malaria is common.[24] This point was illustrated in a description of the experience of a London-based investment company that spent $1.4 billion on an aluminum smelter in Mozambique. Seven thousand of their employees developed malaria. Thirteen of their non-national employees died. Episodes such as this one are likely to make foreign companies hesitate to make investments in countries where malaria is rampant.

Perhaps the most affecting part of the review's tale of woe was told in the two figures that showed the nations of the world. One portrayed GDP and the other the distribution of malaria. Where malaria was the most rampant, the GDP was low. The reverse was also true: where malaria was rare, the GDP was high. One need not be a statistician to see the inverse relationship between the two.

## Increased Sea Level and Storm Surges

Images of Hurricane Katrina and Superstorm Sandy are still fresh in the minds of many. It is not difficult to recall the dramatic photos of people on their roofs waiting to be rescued and those seeking shelter in the Superdome, which was ill equipped to deal with refugees who had fled from their homes. Worst of all, perhaps, was the fate of those unfortunate patients confined to beds and intensive care units in Memorial Hospital in New Orleans. Many died as it became increasingly impossible to provide them with the care they needed. The most distressing part of the ordeal came when the immediate crisis had ended. There were allegations that some patients were euthanized when care options ended. As a healthcare provider, I hope never to be confronted by the situations faced by those who struggled to provide care to patients trapped in this crippled hospital. There were no good options, as revealed by Pulitzer Prize–winning author Sheri Fink in her gripping account, *Five days at Memorial: Life and Death in a Storm-Ravaged Hospital.* (Note: Dr. Fink's Pulitzer was awarded for her equally fine reporting of the Ebola epidemic in West Africa.)

Reports of the economics of sea level rise are usually combined with those of storm surges. There are at least four separate components of the total rise in sea level, as described in chapter 6. It is often difficult or impossible to separate the effects of storm bulges, wind effects on sea level, tides, and wave action. This host of variables makes modeling very difficult. A worst-case scenario would occur if the landfall of a storm coincided with an astronomical high tide in a coastal area that amplifies wind and pressure components—and this essentially is what happened when Superstorm Sandy made landfall. At that time, winds were not particularly strong, but other factors led to a storm surge that was some thirteen feet above a typical low tide mark. This surge was enough to flood the New York subway system and other elements of the New York infrastructure, as described in chapter 6. For virtually all modeling treatments, only sea level rises and storm surges are considered.

Many factors related to changes in sea level exist that are theoretically easier to measure than the effects of climate change on heat-related illnesses. Accurate maps of coastal areas show elevations above sea level. Maps have been entered into global information systems to facilitate computerized analyses. With few exceptions, these elevations are relatively stable. However, in some regions, pumping water from underground sources along with the extraction of oil and gas are causing the land to sink, as discussed in chapter 6. Subsidence is a phenomenon that is equivalent to a rise in sea level. For much of the developed world, there are property assessments that show the value of specific structures at any given location. Thus, the impact and hence the cost of rising sea levels with or without a consideration of storm surges varies widely by region and even by specific localities. Variability and disagreement among sources depends greatly on how much sea level will rise by a given date relative to historical baseline levels.

Sea levels have already risen and will continue to rise even in the extremely unlikely event that virtually all greenhouse gas emissions are curtailed sharply and instantaneously. Figure 6.1 shows tide gauge data from geologically stable locations and data from satellites that also mea-sure sea level. Geologically stable locations are those where there has not been any subsidence or elevation of the earth's crust at the site. Even under the optimistic representative concentration pathway 2.6 (RCP2.6) scenario, warming will continue and its effects will progress—just not as far and fast as in the other, more likely scenarios. As discussed in chapter 6, a substantial portion of the expected rise in sea level will occur as the warmed air transfers heat to oceans as they move toward an equilibrium condition. Warming water expands, and so sea levels will rise. This fact is depicted in figure 6.3, along with other contributors to rising sea level.

Three general strategies may be adopted to mitigate the effects of rising sea level: (1) make an accommodation to the receding shoreline, a process referred to as *nourishment*; (2) build dikes or similar structures to keep the sea in place; and (3) abandon property before it is engulfed. A fourth option—and not a good one—is to do nothing, which is likely to be the costliest option of the four.

In a business-as-usual future, approximated by the RCP8.5 scenario, mean sea levels are expected to increase by 0.6 to 1.7 feet by midcentury and by as much as 4.4 feet by the end of the century.[25] Expected increases in sea level for several major coastal cities are shown in table 9.1.

A suite of expected cost outcomes was published almost a decade ago for different scenarios in which the authors made estimates based on detailed mathematical models. In the study, they assumed that sea level would increase in ten-centimeter increments and that time would advance in ten-year periods. Their model started in the year 2010 and ended in 2100.[26] This setup produced a 9 × 10 table with ten ten-year intervals and

**Table 9.1**

Projected sea level increases

| City | Midcentury rise (feet) | End-of-century rise (feet) |
|---|---|---|
| New York, New York | 0.9 to 1.6 | 2.1 to 4.2 |
| Atlantic City, New Jersey | 1.0 to 1.8 | 2.4 to 4.5 |
| Boston, Massachusetts | 0.8 to 1.6 | 2.0 to 4.0 |
| Portland, Maine | 0.7 to 1.4 | 1.7 to 3.8 |
| Norfolk, Virginia | 1.1 to 1.7 | 2.5 to 4.4 |
| Texas coast | 1.5 to 2.0 | 3.2 to 4.9 |
| Seattle, Washington | 0.6 to 1.0 | 1.6 to 3.0 |
| San Diego, California | | 1.9 to 3.4 |

*Note:* These data presume a business-as-usual scenario for the drivers of climate change, equivalent to the representative concentration pathway 8.5, which projects global temperature increases of around 2°C by midcentury and around 3.7°C by the end of the century. Sea level data were reproduced with permission from Table A15 from the Technical Appendix in K. Gordon, G. Lewis, and J. Rogers, "Risky Business: The Economic Risks of Climate Change in the United States," *riskybusiness.org*, Risky Business Project, 2015.

nine increments in sea level. The table shows that in 2020, the annual cost of a 10 cm rise in sea level if the United States had done everything possible to adapt to a rise in sea level would lead to a $5.51 million loss, a figure that increases to $382 million in 2100 with a 90 cm sea level rise. (Note: The authors calculated costs in 1990 dollars.) In their analysis, adaptation paid for itself by reducing projected losses by about one-third. As one might expect, their calculations were associated with very large uncertainties: a $2 billion estimate. Ninety percent of the estimates made using this model ranged between $0.2 billion and more than $4.6 billion. Nevertheless, a recurring bottom line emerged from this study: adaptation will result in substantial savings.

These authors extended their analysis and introduced a further complicating factor into the analysis by attempting to account for sulfate aerosols in the atmosphere. Sulfates are produced largely due to burning coal and the emission of sulfur oxides into the atmosphere by volcanoes. Some have proposed deliberate injections of sulfates into the atmosphere to prevent climate change, a strategy that is one possible component of geoengineering or climate intervention (see chapter 10). These investigators reported that if high sulfate emission strategies were encouraged, it might be possible to achieve a 55 percent reduction in losses. However, this strategy is unlikely to occur. Many pollution-control strategies focus on *reducing* sulfate emissions in order to protect health and the environment. This is one of the principal reasons for the adoption of the so-called acid rain program in the 1990 amendments to the Clean Air Act.

A more current study reevaluated the combined effects of storm surges and sea level rise on coastal regions of the United States.[27] The authors of this report divided the total cost of adaptation into categories: the value of property that is abandoned in advance of any immediate threat, presumably because it was not thought to be worth protecting; the cost of what they term *armoring* the coastline; the cost of nourishing the shoreline to make it more resistant to storms; the cost of elevating structures; and residual costs. Elevating structures is a reasonably self-explanatory process: buildings are raised up on stilts or pillars so that the structure is higher and therefore more resistant to damage. Nourishing the shoreline includes procedures such as planting flood-resistant species, building salt marshes, pumping sand onto beaches, and other similar adaptive measures. Armoring the shoreline is not as intuitive or straightforward and includes building seawalls, breakwaters, bulkheads, and barricades of rocks, chunks of concrete, and so on designed to blunt the force of waves. In the scenario that postulates that emissions will continue in the present business-as-usual manner until the year 2100, the authors concluded that for the seventeen regions they evaluated, the cost of abandoning property would be around $120 billion, and armoring the shoreline would cost around $190 billion. Coastline nourishment costs were estimated to be about $140 billion, with residual, unaccounted costs coming in at around $150 billion. Elevation costs were relatively small, at around $10 billion. As expected, the costs of adaptation varied substantially among cities studied. A 3°C temperature increase would cost Miami, Florida, around $130 billion to protect against sea level rise and storm surges and around $50 billion to protect against sea level rise only. Miami is a somewhat unusual case. Not only is it flat, with large portions of the metropolitan area barely above sea level, but the underlying coral-limestone rock is porous and water would seep through the rock quite easily. This seepage makes it impossible to construct sea walls to hold back the ocean.

Environmental Justice and Rising Sea Level

On February 11, 1994, President Clinton signed Executive Order 12898 (www.archives.gov, accessed March 6, 2015). This order, titled "Federal Actions to Address Environmental Justice in Minority Populations and Low-Income Populations," requires the EPA and other federal agencies covered by the order to "make achieving environmental justice part of its mission by identifying and addressing, as appropriate, disproportion-ately high and adverse human health or environmental effects of its programs, policies, and activities on minority populations and low-income populations." As an initial step in satisfying this requirement as it applies to rising sea level, it is necessary to determine whether minorities and low-income populations are likely to bear a disproportionate risk of adverse effects as sea levels rise. As is too often the case, the answer is "yes."

It is necessary to have a valid and quantitative measure of social vul-nerability in order to construct a model that includes vulnerability as an outcome. The authors of a recent study used a multistep approach to this end by computing a *Social Vulnerability Index* (SoVI).[28] They extracted values for twenty-six different demographic characteristics from census tract data and condensed them into a single number, the SoVI number. They normalized the data so that the average SoVI was zero. High posi-tive SoVI values indicate high social vulnerability, and low SoVI values indicate low vulnerability. As expected, poverty made the largest contri-bution to a high level of social vulnerability, and wealth was an important determinant of low vulnerability. Once SoVI values were in hand, the authors superimposed these values on maps of the US coastline to create vulnerability maps.

In an independent set of steps, they retrieved data from a sea level rise property value model for each census tract. If the estimated cost of protecting a tract threatened by rising sea level was greater than the value of the property in the tract, they assigned the tract an *abandonment outcome*. If the cost of protection by armoring or nourishing the tract was less than the value of the property in it, they assigned a *protected by armoring* or *nourishing* outcome.

In a final step, the authors merged these two data sets; that is, they combined the SoVI with the abandon, nourish, or armor choices. The

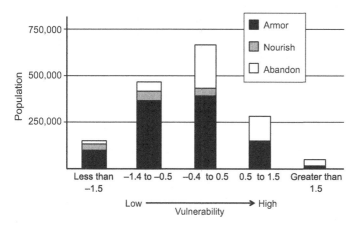

**Figure 9.1**

Adaptation and social vulnerability for the contiguous United States. The total population living in areas that would be abandoned, nourished, or armored after a 66.9 cm sea level rise are shown with their Social Vulnerability Index scores. Moving from left to right along the bar chart demonstrates that as social vulnerability increases, the population protected from sea level rise risk (armored and nourished) decreases, while the population living in abandoned structures increases. As a work product of employees of the federal government, this is not known to be copyrighted.

outcome of this analysis is shown in figure 9.1. The figure shows the number of people living in each of five SoVI categories, arbitrarily grouped for purposes of the analysis, and the property protection strategies dictated by their value. As seen in the figure, the less vulnerable individuals were more likely to live in areas that would be protected. The more vulnerable individuals live in tracts where abandonment is likely. In more quantitative terms, in the Gulf Coast region (Collier County, Florida, to the Texas/Mexico border), 99 percent of the population that is the most socially vulnerable lives in areas that are likely to be abandoned when the level of the sea increases. It is probable that they will become economic refugees. By contrast, 92 percent of individuals with the lowest SoVI, those who are the least vulnerable, live in areas that are likely to be protected from rising sea level.

Although it was not a specific objective of this research, it is likely that socially invulnerable individuals have more political clout than poorer, more vulnerable individuals. These wealthier individuals are almost certain to be better able to influence political decision makers who will determine the response to rising sea level. In addition, the most vulnerable refugees will have the least ability, from a financial perspective, to cope with their new refugee status.

## Agriculture

John Steinbeck's searing novel *The Grapes of Wrath*, published in 1939, describes the fate of the Joad family as they are driven from their land by a combination of drought, the dust bowl, and economic hardship. The scion of this refugee family was portrayed by Henry Fonda in the film adaptation of the novel. Although a reprise of this fictional account is not likely to occur in the United States today, extreme weather continues to affect agricultural production in the Midwest. Between the years 1999 and 2011, a combination of floods, droughts, and other forms of severe weather led to around $90 billion in losses in the agriculture sector of the economy (in 2011 dollars).[29] Heat and drought accounted for much of this loss.

According to the World Bank, agriculture accounted for between 1.2 and 1.4 percent of the value of the gross national product (GNP) in the United States between 2010 and 2012 (http://data.worldbank.org/indicator/NV.AGR.TOTL.ZS, accessed April 22, 2015). Agriculture's contribution to the GNP varied substantially in other nations. Agriculture's contribution to the GNP was greater in the so-called BRIC countries. In Brazil, it contributed around 5.7 percent; in China, the contribution was about 10 percent; and for the Russian Federation, the fraction was 3.9 percent. For India, the contribution was 18 percent in the 2010 to 2014 interval. In some countries, agriculture dominated the GNP. For example, it reached 55 percent in the Central African Republic. It was also dominant in many sub-Saharan nations, where the risk of undernutrition among children is high. As a corollary, in nations where agriculture dominates the economy, the risk to populations posed by crop failures associated with climate change is also high. Failing crops will in turn lead to price increases, undernutrition, and possibly social disruption and violence.

US agriculture produced $470 billion worth of commodities in 2012.[30] In many of the Midwestern and Great Plains states, such as the Dakotas,

Iowa, and Nebraska, agriculture dominates the economy and politics. California produces over 10 percent of the dollar value of US agricultural products and accounts for around half of the fruits and vegetables grown in the country. The threats posed by climate change to agriculture and food production are shown in more detail in chapter 5.

Although earlier springs and warmer temperatures may be beneficial to some plants, high temperatures during critical periods, such as pollination, may cause severe losses. High temperatures need only to persist for a brief time to inflict their damage.

Drought poses additional threats. The megadrought that affected the United States in 2012 cost US farmers an estimated $30 billion. I saw evidence for this myself as I drove through parts of Iowa and Nebraska in the fall of that year. Virtually every field was dry and brown. A World Bank periodical documented a 10 percent increase in the price of food in just one month due to this severe shortage of rain.

To an extent, the increases in the atmospheric $CO_2$ concentration may increase crop yields among the so-called $C_3$ plants, such as wheat and soybeans, while having little effect on $C_4$ plants, such as corn (for more details, see chapter 5). However, the nutritional value of these stimulated plants may suffer.

The authors of the American Climate Prospectus report estimated the effects of climate change on the yields of corn, wheat, and oil seeds under the RCP8.5, RCP4.5, and RCP2.6 climate change scenarios for various portions of the twenty-first century.[31] In their projections, they include the potential effects of $CO_2$ growth enhancement—a factor that is difficult to quantify accurately. Some of the results of this effort are shown in table 9.2. Because the effects become more pronounced toward the end of the century, only those projections for the 2080 to 2099 interval are shown. Similarly, the RCP4.5 scenario is omitted, as those values tend to be between those from the more extreme RCP8.5, business-as-usual, and unlikely RCP2.6 scenarios that would require immediate, huge cutbacks in greenhouse gas emissions. The effects on corn are the most dramatic, because corn is quite heat sensitive and as a $C_4$ plant does not benefit from $CO_2$ fertilization. Under the RCP8.5 scenario, yields are likely to decline by somewhere between 18 and 73 percent. Economic losses would be substantial.

Table 9.2
Impacts of climate change on US agricultural yields with and without $CO_2$ fertilization

| Crop/interval | RCP2.6 | | | RCP8.5 | | |
|---|---|---|---|---|---|---|
| | One in twenty chance less than | Likely | One in twenty chance greater than | One in twenty chance less than | Likely | One in twenty chance greater than |
| Corn | | | | | | |
| 2080–2099 without $CO_2$ effect | -28 | -20 to -0.8 | 1.5 | -87 | -76 to -29 | -22 |
| 2080–2099 with $CO_2$ effect | -27 | -19 to 0.4 | 2.8 | -84 | -73 to -18 | -8.1 |
| Wheat | | | | | | |
| 2080–2099 without $CO_2$ effect | -6.2 | -4.7 to 0.4 | 0.9 | -27 | -20 to -7 | -4 |
| 2080–2099 with $CO_2$ effect | -2.6 | -0.9 to 4.4 | 5.3 | 8.6 | 19 to 50 | 50 |
| Oilseeds | | | | | | |
| 2080–2099 without $CO_2$ effect | -21 | -16 to 2.2 | 4.2 | -82 | -70 to -20 | -14 |
| 2080–2099 with $CO_2$ effect | -18 | -13 to 6.3 | 8.4 | -74 | -56 to 18 | 29 |

Note: Values are for the percent change from 2012 baseline levels. Extracted from the American Climate Prospectus, with permission. Source: T. Houser, R. Kopp, S. M. Hsiang, et al., American Climate Prospectus: Economic Risks in the United States (New York: Rhodium Group, LLC, 2014)

Crime

Violence may become more prevalent as the planet warms, as discussed in chapter 8. According to the Bureau of Justice Statistics, federal, state, and local governments spent around $265 billion on law enforcement and criminal justice activities in fiscal year 2012.[32] Although the authors of the American Climate Prospectus state that crime accounted for the smallest economic impact among the components they studied, this is still a substantial sum and has a high level of visibility, particularly in local television news coverage.[33] Another study predicted that there would be 22,000 murders, 180,000 cases of rape, 1.2 million aggravated assaults, 2.3 million simple assaults, 260,000 robberies, 1.3 million burglaries, 2.2 million cases of larceny, and 580,000 cases of vehicle theft in the United States between 2010 and 2099 that could be attributed to climate change.[34] Because the temperature effect on crime is greatest during hotter summer months, particularly for vehicle theft and larceny (good weather crimes), month-by-month predictions peak during June, July, and August. Vehicle thefts are at their lowest when there is snow on the ground. Murder is an exception to the winter trough, summer peak rule, showing rather flat rates throughout the year. Another study suggests that increases in violent crime during hot spells are followed by a diminution in the rate in the week immediately thereafter.[35] This may be akin to the harvesting effect associated with deaths attributed to heat waves.

In keeping with the premise that hotter weather leads to more crime, under the RCP8.5 business-as-usual scenario the cost to the US public will increase during this century. Between 2020 and 2039, the likely increase in crime-related cost ranges between $0 and $2.9 billion, rising to $1.5 to $5.7 billion between 2040 and 2059, and increasing still further to between $5.0 and $12 billion in the 2080 to 2099 interval. Different states bear different burdens. The average per capita costs are predicted to range between $16 and $37. The highest per capita increases are predicted to occur in Michigan, New Mexico, Maryland, and Illinois, with the lowest in Utah, the New England states, and Washington. Figure 9.2 depicts the increased costs per capita between 2080 and 2099 in 2011 dollars, as predicted by the RCP8.5 scenario conditions.

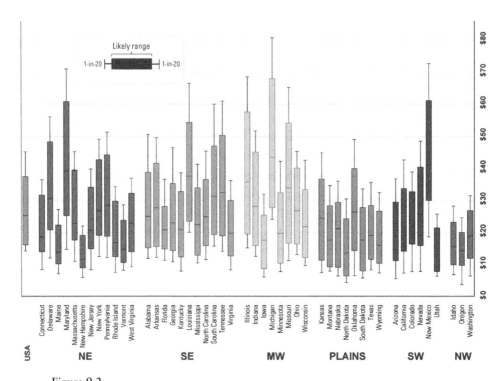

**Figure 9.2**

State-level direct cost increases from changes in crime rates between 2080 and 2099 under conditions predicted by the RCP8.5 climate change scenario. Reproduced with permission from T. Houser, R. Kopp, S. M. Hsiang, et al., *American Climate Prospectus: Economic Risks in the United States*, 109 (New York: Rhodium Group, LLC, 2014).

### The Trajectory toward the Future

The task of assessing the economic costs of climate change on health is one that has no boundaries. The cost to an individual for an occurrence or exacerbation of a specific disease, such as malaria, includes some straightforward aspects. Add up the cost of medications, doctor or clinic visits, and days lost from work, then expand that total by considering how much local or national governments and agencies spend on mosquito control, water drainage, disease monitoring, and so on. Add to this the impact of the condition on the family. Does it change family dynamics, number of children, whether all go to school or work? What is the loss to society? Now substitute a single act of violent crime that occurs during a heat wave for an attack of malaria. Repeat the litany of the expanding ripple effect as an event that seems to be isolated spreads throughout a family and community. Repeat again, again, and again. It is like trying to find the end of a Möbius strip. By definition, there is none.

What does all of this have to do with health? Here, I would remind you that the World Health Organization explicitly includes mental and social well-being in its definition of *health*. The economic vitality of a family, community, state, or nation is one of the critical elements that determines social well-being.

Charting a path toward better financial well-being by mitigating and adapting to climate change will require the same elements outlined at the end of chapter 1, including political leadership and stakeholder involvement.

# 10
## Protecting Health

I know this ... a man got to do what he got to do.
—John Steinbeck, in *The Grapes of Wrath*[1]

Overshoot, adapt and recover. We will probably overshoot our current climate targets [i.e., temperatures will be higher than hoped for], so policies of adaptation and recovery need much more attention.
—Martin Parry, Jason Lowe, and Clair Hanson, commentary in *Nature*, 2009[2]

### A Medical Framework for Climate Change Prevention

*Prevention is better than treatment* is a medical axiom. Healthcare professionals typically think in terms of primary, secondary, and occasionally tertiary prevention of disease. *Primary prevention* refers to the steps needed to prevent the occurrence of a disease. Referring back to table 1.2, hypertension, the leading risk factor for disease throughout the world, serves as an example.[3] The National Heart, Lung, and Blood Institute of the National Institutes of Health (NIH) lists the following primary prevention interventions with documented efficacy: weight loss, restricting dietary sodium, increased physical activity, moderation of alcohol consumption, supplementing the diet with potassium, and consuming a diet that is low in saturated and total fat and rich in fruits, vegetables, and low-fat dairy products.

Primary prevention of climate change requires a reduction in the emission of greenhouse gases. In order to maintain our energy-dependent economy, this means replacing energy sources that depend on burning fossil fuels with energy sources that don't, such as wind, solar, water, and others. It also means supporting energy research that opens other possibilities, such as artificial photosynthesis and sunlight-mediated hydrolysis of water.

Secondary prevention requires taking steps to prevent recurrence or progression of a disease once it has been diagnosed. For hypertension, this usually means starting drug treatments under a physician's supervision. Low-dose enteric-coated aspirin is used widely for secondary prevention of stroke, with the goal to prevent a second stroke. In a climate change context, secondary prevention means taking the steps necessary to reduce threats that are on the horizon. In other words, it is necessary to adapt to the changing climate by taking action to prevent additional damage. Adaptation may include mosquito control measures, building sea walls, developing drought- and heat-resistant crops, strengthening the public health infrastructure, and many others.

Tertiary prevention involves helping patients manage long-term health problems to maximize quality of life. These measures may include rehabilitation programs and support groups. Often, the lines between secondary and tertiary prevention blur under the rubric of climate change. Adaptation is at the center of both.

Many sectors of society should and must be involved in preventive measures that center on climate change.[4] These include international organizations, such as the United Nations; national governments, particularly in countries that emit large amounts of greenhouse gases; state and local governmental agencies; nongovernmental organizations, such as Physicians for Social Responsibility, Natural Resources Defense Council, the Sierra Club, and Earthjustice, to name but a few; research universities; national laboratories; and the private sector. Although individual concerned citizens may feel helpless when working to prevent climate change, everyone can make a contribution. We are all stakeholders, and stakeholders can and must be involved.

Just as conventional medicine struggles to deal with any severe medical problem, society needs as many strategies as possible to deal with climate change in order to minimize the health and environmental impacts that are so clearly on the horizon.

The IPCC authors conceptualized relative risks to health due to climate change in a series of target-like pie graphs, as shown in figure 10.1.[5] The size of each slice of the pie is proportional to the risk and the potential for risk reduction. These judgments were based on a critical

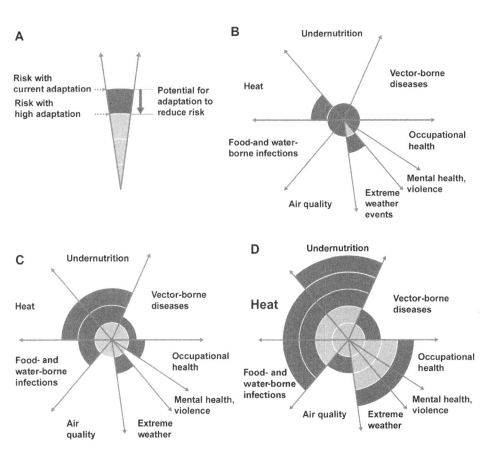

Figure 10.1

Targeting health. Panel A is a key to understanding the remainder of the figure. The magnitude of the risk for each factor is shown by the width of the slice of the pie. The darkened portion of each slice depicts the potential for risk reduction in a hypothetical, highly adapted condition. Panel B shows the relative importance of the burden of poor health at present in a qualitative way. Panel C depicts the risks of and potential benefits to be gained from adaptation in the relative near term, 2030 to 2040. Panel D portrays the relative risks and adaptation potentials toward the end of the century, 2080 to 2100, with a temperature rise of 4°C relative to the preindustrial era. *Source:* Figure 11–6 (originally in color) from D. Campbell-Lendrum, D. Chadee, Y. Honda, et al., "Human Health: Impacts, Adaptation, and Co-Benefits," in *Climate Change 2014: Impacts, Adaptation, and Vulnerability; Part A: Global and Sectoral Aspects; Contribution of Working Group II to the Fifth Assessment Report of the Intergovernmental Panel on Climate Change,* ed. C. B. Field, V. R. Barros, D. J. Dokken, et al., 709–754 (New York: Cambridge University Press, 2014). Reproduced per IPCC guidelines.

evaluation of the literature and the judgment of the authors. At present, risks are real but relatively small compared to risks posed by future temperature increases. As temperatures increase, risks rise. There are reasons to be hopeful. As shown in the figure, risks can be reduced substantially even if temperatures rise by as much as 4°C, but only if adaptation measures are intense.

Parry, Lowe, and Hanson, the authors of the second quotation at the beginning of this chapter, made several important points that emphasize the reality of the situation we face. They wrote that those engaged in efforts to halt climate change by reducing the emission of greenhouse gases "need to be optimistic that their decisions could have swift and overwhelmingly positive effects on climate change."[6] Although these words were written in 2009, the message is still true and important today. The climate is changing slowly, and the benefits of mitigation and adaptation will not be evident until far into the future. The authors gave several examples: they stated that if emissions were to peak in 2015, temperatures would not peak until 2065 (a peak-to-peak interval of fifty years). Delaying the emissions peak by ten years, to 2025, pushes the temperature peak an additional fifteen years into the future, to 2080 (a peak-to-peak interval of fifty-five years). In their most extreme scenario, if the emissions peak did not occur until 2035, the ever-rising temperature peak would not occur until 2100 (a peak-to-peak interval of sixty-five years). Do the math: deferring the emissions peak by twenty years delays the temperature peak by sixty-five years (2100 − 2035 = 65). This is around the average lifetime for most citizens of the world.

In 2008, the United Nations Framework Convention on Climate Change updated its earlier report on the investments needed to mitigate and minimize climate change.[7] They estimated that expenditures of around $86 billion would be needed in 2015. In its update of the 2007 report, the committee estimated that an additional annual investment of between $200 and $210 billion would be needed to reduce the emission of carbon dioxide equivalents to levels 25 percent below 2000 levels by 2030. This level of funding is nowhere in sight. To paraphrase the Steinbeck quote, "What does a man got to do?" That which needs to be done varies substantially among countries and governmental entities.

## Adaptation

### Heat

There are limits to everything. This is particularly true for the ability of humans to withstand elevated temperatures. Relatively sharp limits define the range of temperatures within which humans can survive.[8] At rest, all adults and children generate heat as the result of normal metabolic processes. There is a minimum amount of heat that cannot be reduced. Normally, we generate about one hundred watts of power in a resting state. Naturally, any increase in activity results in an expenditure of energy that increases heat output as we do work. This increase can be very large under conditions of extreme exercise. In addition, wearing bulky clothing or uniforms (e.g., football uniforms) interferes with the loss of heat needed to maintain a normal body temperature, which causes the retention of body heat and an increase in body temperature. This is why football players and others are particularly susceptible to developing heat-related illnesses during practices or games: they wear bulky protective gear and engage in very intense exercise.

In order to maintain a normal body temperature of 98.6°F (37°C), the body must be able to dissipate enough heat to prevent an increase in body temperature. We dissipate heat, regardless of how it is generated, by four mechanisms: convection, conduction, radiation, and evaporation. Evaporative heat loss is the most important of these and mostly occurs via sweating. It is also possible to dissipate heat by immersion into water, a more efficient but often impractical alternative to the usual process of sweating. Ambient temperature and humidity limit the ability to lose heat by evaporation. Above a critical temperature, the body gains heat no matter what. This limit is typically defined by what is known as the *wet bulb temperature* ($T_W$), the temperature registered by a well-ventilated ordinary thermometer covered by a wet cloth. The $T_W$ limit for humans is around 35°C with a relative humidity of 100 percent. Hyperthermia develops at temperatures above this point and can lead to heat exhaustion or to the more serious and potentially fatal condition of heat stroke, as discussed in chapter 3. As the climate warms and more water enters the atmosphere from warming oceans, temperatures that exceed the $T_w$ threshold will become more common. Adaptive measures will become increasingly important.

The Chicago heat wave of July 1995 was a learning experience for the public health community. An analysis of the Chicago data helps point the way toward the development of adaptive measures that should limit morbidity and mortality associated with heat. The most important of these measures is improving access to air conditioning. Public health programs that identify those at risk, reach out to them at a time of an impending heat emergency, and move them to safer locations should be protective.

Strategies that have a longer lead time and are designed to alter the space in which we live, work, and play on a daily basis (the built environment) should be helpful from any one of several perspectives. Roofs that are white reflect more heat than those that are black (common in tar and gravel roofs). Buildings with "green roofs" that are covered by grass or other plants will benefit their occupants. These measures will lead to cooler buildings, particularly those that are poorly insulated. Cooler buildings require less air conditioning to maintain healthy indoor temperatures. Reducing the need for air conditioning also reduces the amount of indoor heat transferred to the environment, which attacks the heat island effect that is increasingly common and problematic in modern urban areas. Lowering the air conditioning demand also reduces the consumption of electricity, particularly at times of peak usage, and the need to burn fossil fuels to generate that electricity.

### Chicago's Climate Action Plan: A Blueprint for the Future

Spurred on by the realities of climate change and a strong desire to avoid the consequences of another heat wave such as the one that devastated the city in 1995, Chicago established a task force to plan for the future. This task force has done more than kick the can down the road a few feet: it has created a series of steps that are both ambitious—but workable, in the eyes of its creators—and necessary to protect the health of Chicagoans and designed to bring the city and its residents safely and responsibly into the next century. Details of the plan can be viewed at www.chicagoclimateaction.org.

The plan's creators began by confronting some of the realities of their present situation and likely scenarios for the future. The predictions by climate change experts suggest Chicago will have a climate like that of Mobile, Alabama, by the end of the century. The number of days when the temperature will exceed 100°F will rise from two per year to thirty-one. An energy and greenhouse gas inventory showed that the city emitted around 36.2 million metric tonnes of carbon dioxide equivalents each year (MMTCO$_{2e}$; one metric tonne = 2,200 lb.). That works out to be around 12.7 tonnes per person. In keeping with the Kyoto Protocol, the plan sets a target of a 25 percent reduction in CO$_{2e}$ emissions by 2020 and an 80 percent reduction by 2050. The plan to achieve this goal focuses on five objectives:

- Improving the energy efficiency of buildings
- Using clean renewable sources of energy
- Improving transportation options
- Reducing waste and industrial pollution and improving energy efficiency
- Adapting to higher temperatures

The task force came to the realization that most of the readily achievable goals were associated with buildings and transportation. Around 70 percent of the city's greenhouse gas emissions were attributed to buildings and 21 percent to the city's various forms of transportation—mostly from cars, trucks, and buses. Only 9 percent were the result of the combined effects of all other sources.

A strategy to retrofit existing buildings with more energy-efficient systems is central to Chicago's plan. In one example that the plan creators cite, F&F Foods spent $780,000 on an energy-efficiency project that led to an immediate savings of $280,000 per year. The payback time was relatively short: 2.6 years. Similar retrofitting projects applied to half of the city's commercial and industrial buildings are predicted to yield a 30 percent savings in energy, equivalent to 1.3 MMTCO$_{2e}$ annually. A 50 percent improvement in the energy efficiency of residential buildings would translate into another 1.44 MMTCO$_{2e}$ per year. Replacing tungsten light bulbs with compact fluorescents or light-emitting diodes (LEDs) and exchanging old, energy-hogging appliances such as older refrigerators with more modern, efficient models would yield another 0.28 MMTCO$_{2e}$. As a concrete example, my wife and I recently replaced a refrigerator/freezer purchased in 1992 that had an annual energy consumption of 1600 kWh of electricity (per the EPA) with one that uses about 25 percent of that amount of electricity. Updating the Chicago energy code, establishing new guidelines for building renovations, building more cooling parks, lining streets with trees, and creating green roofs would save another estimated 1.61 MMTCO$_{2e}$ each year. Other steps listed would yield an additional 0.84 MMTCO$_{2e}$.

Substantial savings are available by improving the electrical power supply for Chicago. By building renewable energy sources for the city, the task force estimates that emissions from that sector can be reduced by 20 percent, or 3.0 MMTCO$_{2e}$ per annum. Upgrading power plants (specifically, twenty-one located in Illinois) and making other improvements in the efficiency of existing power plants would yield an estimated 3.54 MMTCO$_{2e}$.

Although utilities claim that they encourage improvements in the energy supply system, some promote policies that discourage or stop homeowners from net metering. *Net metering* allows your electric meter to run backward if your home systems generate more electricity than you are consuming at that moment. For example, an Arizona utility, located in the state that has the largest potential for capturing solar energy, proposed imposing fees that would make rooftop photovoltaic electricity prohibitively expensive for individuals. Similar threats are arising elsewhere.

Chicago is currently one of the hubs for rail transportation. It is also the site of congestion that paralyzes the rail industry. A 2012 article in the NY Times reported that a train that travels from Los Angeles to Chicago can make the journey in forty-eight hours—then it may take an additional thirty hours to pass through Chicago. The reporter followed a train full of sulfur. It took twenty-seven hours to pass through the city at an average speed of 1.13 miles per hour. This pace is slower, or so the reporter wrote, than that of an electric wheelchair![9] Improving freight movement was estimated to reduce annual emissions by 1.61 MMT-CO$_{2e}$. Investing in more public transit would save 0.83 MMTCO$_{2e}$. Other transportation-related improvements were identified, including transit-oriented development, making walking and bicycling easier (a health-promoting goal in and of itself), carpooling and sharing, replacing fuel-inefficient cars and trucks with those with higher fuel efficiencies (already a part of the corporate average fuel economy standards agreed to by the motor vehicle industry and the EPA), using cleaner fuels, and other strategies. These strategies would raise the annual total energy savings to over 2.5 MMTCO$_{2e}$. Mitigation strategies involving reduce, reuse, recycle promotions and switching to refrigerants that are not greenhouse gases would add another 2.1 MTTCO$_{2e}$ to the total.

Finally, Chicago will update the heat response plans for the city to focus on assisting vulnerable populations. The plan creators hope to make the city's urban areas greener, to preserve trees and plants, and to make cooling more innovative. Improvements in power plant design and operation will reduce the concentration of ground-level ozone, protect air quality, and promote better health. Again, children, the elderly, and those with chronic diseases would benefit the most.

These ambitious goals need support at all levels. Ordinary citizens, city officials, and stakeholders must join forces to combat the opposition that is certain to arise and to be well financed. Preventing climate change will save energy. By the end of the century, a low emissions scenario could limit the number of days in Chicago for which temperatures exceed 100°F to six, compared to the thirty-one days predicted by a business-as-usual scenario. This reduction alone will save enormous amounts of electrical energy, resulting in fewer emissions of greenhouse gases and better health.

This example of a city working to reduce carbon dioxide emissions, save energy, and improve health represents one strategy for moving forward toward a more sustainable energy future and a future that holds the promise of better health for its citizens. There are other paths forward, but they require political leadership, stakeholder involvement, and other elements (as outlined in chapter 1).

## Infectious Diseases

A variety of tools are available to combat the threats posed by infectious diseases as the climate warms. They range from the simple to the complex—from hand-painted signs to satellites in orbit. All have a role to play.

Multiple effective strategies are available to prevent mosquito-borne illnesses. It is important to prevent bites by infected mosquitoes. When possible, people at risk should remain indoors during dawn and dusk hours when mosquitoes are on the prowl for their next meal. Wear protective clothing. Use insect repellants containing N,N-diethyl-meta-toluamide. (No wonder this chemical is commonly called DEET!) When appropriate, sleep under insecticide-impregnated nets to keep insects from biting while sleeping. Make certain that there are no open containers around that will store or trap water, providing potential breeding sites for mosquitoes. Although effective, these measures should be supplemented by other, more aggressive and targeted measures designed to keep mosquitoes from reproducing.

## Malaria

Combating malaria was chosen as one of the Millennium Development Goals because the magnitude of the problem is high and the effect on children is disproportionate. In spite of the progress described ahead, malaria must remain a high-priority target in the task of adapting to climate change.

The most recent World Malaria Report, from 2014, makes it clear that the official estimates of the number of malaria cases as presented in the report are inadequate.[10] The WHO numbers include just those individuals with what they call "patent" infections—that is, those with evidence of an infection found on light microscopy of a stained blood smear or a laboratory diagnostic test. Even with that limitation, the number of infections is staggering. The eighteen sub-Saharan nations account for an estimated 90 percent of the infections in that vulnerable part of the world. Nigerians harbor an estimated thirty-seven million infections and the Democratic Republic of the Congo another fourteen million. The World Malaria Report indicates that the number of Africans who have low-intensity infections is considerably higher. In fact, one might argue that virtually all Nigerians, particularly those who live in rural areas, are infected at some time during their lives.

The 2014 World Malaria Report contains good news and bad news— and some of that news is or ought to be embarrassing.[11] The good news is that since 2000 there has been a sharp reduction in the malaria prevalence among children between two and ten years of age. In 2000, 26 percent of children were infected, a number that fell to 14 percent in 2013. Correspondingly, the number of malaria infections dropped from 173 million to 128 million during that same period. Finally, malaria mortality rates fell by 47 percent worldwide and even more, by 54 percent, in the WHO-designated Africa region. Given the circumstances in effect at the time of the report, all indications are that there will continue to be progress in the control of this disease, one of the primary goals set forth in the Millennium Development Goals. The authors of the report project a 55 percent reduction worldwide and a 62 percent reduction in the WHO Africa region if the preceding thirteen-year rates hold through 2015. Additional progress toward the reduction of mortality among children less than five years old is expected, with a 61 percent reduction globally and a 67 percent reduction in the WHO Africa region.

The bad news in the report is that much more could be done. Only $2.7 billion of the $5.1 billion, or just under 53 percent of the funds that would be likely to achieve worldwide control of malaria, were made available from international and domestic sources. Funds are needed to finance steps that are proven to work:

- Supplying enough insecticide-treated bed nets (ITNs) for populations at risk, particularly children
- Providing additional protection for pregnant women by treating them with what is referred to as *intermittent preventive treatment in pregnancy* (IPT$_p$)
- Increasing the availability of definitive testing for malaria and distributing so-called artemisinin-based combination therapy (ACT)

ITNs are a mainstay in the prevention of malaria. These nets operate by several mechanisms. First, they kill and repel mosquitoes, thereby protecting people from bites of the disease-carrying females while they sleep. Second, there is an aspect of "herd immunity" that occurs when enough people in a house or community are protected by nets. The repellant and insecticidal properties provide an element of protection to those not under an appropriate net. The CDC's malaria website is an excellent ITN reference.[12]

The first nets developed were treated with *pyrethroids*, members of a class of insecticides that is thought to have relatively few adverse effects on the humans it is designed to protect. This is especially true when compared to other insecticides, particularly the organophosphates. The *organophosphates* are close relatives of nerve gases and have many toxic effects. Nets treated with pyrethroids do have a distinct disadvantage: they do not last very long. In addition, exposure to sunlight ruins their ability to kill and repel mosquitoes. Therefore, to remain effective, they must be periodically immersed in a water–pyrethroid solution and allowed to dry in the shade. The typical useful lifetime of these nets is about a year. Nets that last longer are clearly desirable. As of February 2014, the WHO gave full or interim approval to eleven more desirable, long-lasting insecticide-treated nets, or LLINs.[13] These nets typically remain effective for around three years. The CDC website suggests that it would be possible to realize a savings of $3.5 billion over ten years by extending the useful lifetime of LLINs from three to five years.

The immunity that appears to develop in adults who are routinely and repeatedly infected by the malaria parasite may wane during pregnancy. The cause of this partial immunological failure has been attributed to the changes in the pregnant woman's immune system and the presence of a new organ, the placenta, which is a potential binding site for the malaria protozoans. The CDC and WHO websites both recommend that a curative dose of sulfadoxine/pyramethamine (SP) be given to all pregnant women whether an infection is present or not as part of IPT$_p$. Folic acid is also given to women to prevent neural tube defects, such as spina bifida. However, in high doses folic acid inhibits the action of SP. Therefore, the CDC recommends a folic acid dose of 0.4 mg or less. In places where a 5 mg dose is routine, SP treatment should be suspended for two weeks after the vitamin is administered. Unfortunately, this makes compliance less likely. SP therapy has a number of benefits associated with the prevention of a malaria infection, including reducing the risk of premature delivery, preventing intrauterine growth retardation, and reducing the probability of delivering low birth-weight babies (less than 5.5 pounds or 2.5 kg); preventing fetal loss; and reducing the risk of anemia in the mother. IPT$_p$ should be given in addition to the routine use of ITNs (or LLINs). Women should be monitored during pregnancy, and effective treatment should be administered if malaria is diagnosed before the delivery of a baby.

Although fever is a common symptom among individuals with malaria, there are many other infections that are heralded by an increase in temperature. Most tragically, this was seen during the Ebola epidemic that peaked between 2014 and 2015 in West Africa. Ebola overwhelmed many clinics that were then unable to treat other patients. Also, fear of Ebola kept many febrile individuals from receiving proper diagnostic testing. As a result, many patients who had malaria or other treatable diseases went undiagnosed and untreated. That specific issue aside, the universal availability of rapid testing either by microscopic examination of the blood or a rapid serological test for malaria coupled with instant access to treatment would reduce malaria-associated morbidity and mortality substantially. Treatment for uncomplicated malaria due to *P. falciparum* is thought to reduce the mortality in children between the ages of one and twenty-three months by as much as 99 percent. The results are almost as good in older children less than five years of age.

The availability of rapid diagnostic tests (RDT) for malaria has increased dramatically in the past decade. The World Malaria Report puts the number of tests distributed by malaria control programs in 2013 at 160 million, up from two hundred thousand in 2005.[14] In parallel with this increase, the reliability of the tests has also increased. Manufacturers of RDTs reported selling almost 320 million tests in 2013. Of these, around 60 percent were specific for *P. falciparum*. The remainder were so-called combination tests that are able to detect more than one species of the malaria parasite.[15] The goal of producing an effective vaccine for malaria remains elusive, although progress has been made. Producing an effective vaccine against protozoal diseases is difficult for many reasons. The life cycle of the parasite is complex, as discussed in chapter 4. Each stage of the infection poses different challenges for those who are working to develop an effective vaccine.

In the fall of 2011, the first results were published from a large clinical trial in which a malaria vaccine was compared to a nonmalarial vaccine.[16] This report described the results from the first six thousand children entered into the study. The group consisted of children who were between the ages of five and seventeen months at the time that the first of three doses was administered and who completed all three of the scheduled immunizations. Researchers also studied another group of younger children between the ages of six and fifteen weeks. In the older group, the incidence of clinically proven malaria was 0.32 episodes per person per year in the vaccinated group compared to 0.55 episodes per person per year in the control group. Efficacy in the combined age group was around 35 percent. Serious adverse events were comparable in all groups, including those treated with the placebo. The investigators noted that during the course of the study they transported participants to study sites so they would not miss an injection. This level of support is much less likely to occur in a real-life situation. Although the results are promising, there is

still considerable room for improvement. If the vaccine had been spec-
tacularly successful, it is probable that the safety monitoring committee
would have stopped the study, as continuing to offer placebo vaccination
would have been deemed unethical. This did not happen. It is likely that
the committee decided that the most ethical choice was to continue the
study as designed.

As time passes and as public health officials develop effective partner-
ships with climatologists with access to satellite-based data, it is reasonable to expect that it will be possible to deploy resources needed to
prevent malaria more effectively. This is illustrated, as described in chapter 4, by using climate predictions based on El Niño and the atmospheric
pressure correlate, the Southern Oscillation, to predict periods of higher
risk of malaria in Botswana.[17]

More robust public health infrastructures would make it much easier
to meet the WHO's Millennium Development Goals for malaria in developing countries. This is particularly true for Africa. The barriers to achieving that objective are daunting—and poverty may be the biggest. As
discussed previously, malaria and poverty are locked in a destructive positive feedback cycle; either element makes the other worse. Weak, corrupt
central governments also make bad situations worse, and weak governments also make it possible for terrorist organizations such as Boko
Haram to proliferate, gain strength, and disrupt existing public health
infrastructures.

The quote by Margaret Chan, the director general of the WHO, that
introduces chapter 4 points out the precarious nature of the progress
made toward the control of malaria. In the 2013 World Malaria Report,
she wrote that "the great progress that has been achieved could be undone
in some places in a single transmission season."[18]

## Dengue

In February 2015, my wife and I visited Siem Reap, Cambodia. This city
is well-known for its proximity to Angkor Wat and other spectacular
temples from the days of the Khmer Empire. Visiting these temples
was the main purpose of our trip. Our tour company advised us that
"medical facilities and services in Cambodia do not meet international
standards." We knew that dengue is endemic in Cambodia. This was
reinforced by one of our guides, who told us that one of his children had

the disease. We carried lots of 100 percent DEET mosquito repellant. Although we were forewarned, we were surprised to see a street sign asking for blood donations in front of a pediatric hospital. Dr. Beat Richner, the founder and primary supporter of the hospital, named it after Jayavarman VII, who reigned between 1181 and 1218 CE. He was perhaps the greatest king of the Khmer era, and he built more than one hundred hospitals. Dr. Richner is an accomplished cellist who performs frequently in Cambodia and his native Switzerland to raise money for the hospital. A photograph of the sign pleading for blood donations to treat children with dengue hemorrhagic fever is shown in figure 10.2. It is in English

**Figure 10.2**

Plea for blood donations to treat children with dengue hemorrhagic fever. This sign is in front of the Jayavarman VII Hospital in Siem Reap, Cambodia. It is in English and clearly aimed at the increasing number of tourists who visit Angkor Wat and other Khmer temples that surround the city.

and clearly aimed at the huge number of tourists who flock to the temples (for good reason).

Developing, testing, and implementing measures designed to adapt to the realities posed by this complicated disease will be difficult. One strategy for controlling dengue depends on the elimination of breeding sites for *Aedes aegypti*, the mosquito vector for the disease. The determination of how rigorous a breeding site elimination program must be depends on a number of variables, including the number of people in the population who have already been exposed to the virus (the seroprevalence, which ranges from 0 to 100 percent), the temperature of the air in the region under consideration, and the number of mosquito pupae per person. The authors of a study from 2000 determined that if between 0 and 67 percent of the population had antibodies to the dengue virus in its blood, the threshold level for infection ranged between 0.5 and 1.5 *Ae. aegypti* pupae per person. The results of the investigations were discouraging. In places like Mayaguez, Puerto Rico, the researchers concluded that seven out of every seventeen standing water containers must be eliminated for the prophylaxis program to succeed. In other places, such as Trinidad, twenty-four of twenty-five breeding sites would need to be eliminated.[19]

In order to determine how to best focus *Ae. aegypti* eradication efforts, a group of investigators based in Trinidad inspected over 1,500 homes containing over twenty-four thousand containers.[20] In these homes, 223 harbored larvae or pupae of the offending mosquito. The investigators classified forty-one of these as *key premises* or key locations because they persistently tested positive for this dengue vector. An analysis of which container type was most likely to harbor the vector was revealing. Water drums were at the top of the list, with 53.5 percent containing the larvae or pupae. Next came buckets, with 22.2 percent containing larvae. Then, 8 percent of tubs and basins were infested. This was followed by water tanks, which accounted for 5.4 percent. Tires came last, accounting for just 2 percent. The authors argued that better control of the vector would be achieved if mosquito eradication efforts were focused on treating or emptying the high-risk containers in the key premises. To maximize the effectiveness of interventions, eradication efforts should be focused on months with the highest rainfall, when dengue cases are most likely to occur.[21]

About two decades ago, another study addressed issues that arose as the result of a loss of effectiveness of what was termed *ultra-low-volume insecticide spraying*.[22] This spraying was used to control the population of adult mosquitoes. The study authors concluded that a combination of community-based and centralized approaches, vertically structured so that a central authority controlled most elements of the eradication program, offered the best chance of success. This finding speaks to the over-arching importance of developing and sustaining stable governments and a vigorous and effective public health infrastructure.

Because dengue is a viral disease, there is hope that a vaccine will be developed to prevent infections, reduce morbidity, and save lives. The results of a trial among children were published in early 2015.[23] Because dengue was endemic in the region where the vaccine was being tested, a subset of the children were tested to determine how many had evidence of a prior infection by any of the four subtypes of the virus (see chapter 4 for additional information about the virus). It was no great surprise that almost 80 percent of the children had evidence of a prior infection. This made it possible, or even likely, that children with evidence for a prior infection might be reinfected with one of the other subtypes of the virus.

Because many children had evidence of a prior dengue infection, it was necessary to enroll a large number of participants in the trial to obtain a statistically valid result. This study included over twenty thousand children between the ages of nine and sixteen who lived in one of five Latin American countries. Because the trial was so important, it merited publication in *The New England Journal of Medicine*, perhaps the leading US medical journal. The trial was designed to determine whether the vaccine worked; that is, it was an efficacy trial, designed to determine whether vaccine would protect children from dengue. It was not known whether the vaccine actually worked, so it was necessary to administer a placebo vaccine to some participants. This choice is ethical when the efficacy of a treatment is not known. Two-thirds of the children were randomly assigned to get the vaccine, and the other third received the placebo, an approach that anticipated benefit but also ensured that the trial would be ethical and provide a definitive answer. The test substance (vaccine or placebo) was given at months zero, six, and twelve, and the children were followed for twenty-five months. To avoid loss of the participants, they were contacted weekly to see whether they were well. Since there are many reasons why the children in the study might get sick (as any parent knows), children were seen by a member of the study team within five days of developing a fever. Blood tests were performed at that time and again one or two weeks later to determine whether the child had dengue or some other illness.

In the final analysis, it was shown clearly that the vaccine worked, but it worked better against some types of the virus than others. Overall, the efficacy was just barely under 65 percent for children who received at least one of the three planned doses of the vaccine. Efficacy was about 50 percent for serotype 1, 42 percent for serotype 2, 74 percent for serotype 3, and just under 78 percent for serotype 4. Those who contracted dengue in spite of their vaccination may have had partial protection, as the efficacy against hospitalization was about 80 percent. The vaccine was 99.5 percent effective for preventing dengue hemorrhagic fever, the most severe form of the disease. The prevalence of adverse events was no higher among the vaccinated children than among those who received the placebo—additional good news. Because the children were followed for just twenty-five months after they were vaccinated, the study could not determine how long the vaccine will continue to provide protection. This study confirmed an earlier Asian investigation that was slightly smaller and funded by the same pharmaceutical company.[24]

Taken together, these two studies illustrate the difficulties encountered when attempting to determine whether a vaccine works and is safe. Neither study specified the cost, but the costs were almost certain to have been very high. However, the benefits almost surely will be enormous. Remember, with the business-as-usual climate scenario, about half of the world's population and a great deal of the US population will be at risk by the end of the century (for additional details, see chapter 4). An effective vaccine will be most welcome.

### Agriculture, Food Production, and Food Security

More than 70 percent of all agricultural systems depend on the amount of rain that falls on croplands.[25] This fact makes food production and food security highly vulnerable to climate change—perhaps more vulnerable than other segments of the economy. If there is too little rain, crops fail because of drought. If there is too much rain, crops will also fail because they will drown or they cannot be planted because fields are inaccessible. Chapter 5 reviewed some of the other factors that contribute to the sensitivity of agriculture to climate change. Temperature, the atmospheric concentrations of $CO_2$ and ozone, the impact of plant diseases, and the proliferation of weeds are all related to the climate and will affect agricultural productivity.

Although climate change is likely to benefit agricultural productivity in some regions, it may be devastating in others. For example, growers in the high latitudes may benefit from longer growing seasons and the possibility that they may be able to plant and harvest two crops instead of just one. However, in the low latitudes nearer the equator, climate change may lead to severe crop failures or the inability to grow much at all. Unfortunately, it is these low-latitude nations—such as sub-Saharan Africa and India—that already suffer from food insecurity. Substantial numbers of children in these areas are undernourished—and they will suffer the most.

The yields of some crops will increase in a warmer world. IPCC estimates suggest that around 10 percent of crops will have yield increases of around 10 percent. For another 10 percent of the total yield, losses of 25 percent can be anticipated. Overall, losses will be greater than gains. The IPCC authors predict that in the absence of adaptation a temperature increase of 2°C will reduce the yields of major food crops, such as rice, wheat, and corn. After the middle of the century, things are likely to become even worse.

Although it is likely that there is a limit to successful adaptations to climate change, it is essential to apply measures that are known to work as widely as possible. These measures include

- changes in planting practices;
- alterations in harvesting;
- adapting fertilizer and water usage practices—that is, harvesting rainwater for use in agriculture;
- increasing crop diversity—because monoculture agriculture risks massive failure if the planted crop fails to resist stresses imposed by changes in rainfall, pest proliferation (including invasive weeds), and so on; and
- improving conservation methods to better protect existing resources.

## Agricultural Adaptation

Agricultural practices vary enormously within the United States and even more across the globe. Thus, it is no surprise that the strategies needed to cope with the changes that are already occurring must also vary. Different crops, growing conditions, and forecasts for the future will all require different adaptations if worldwide agriculture is to keep pace with a growing population. Many of the details associated with the methods needed for agriculture to adapt successfully to climate change are beyond the scope of this book. However, reviewing several examples of successful adaptation in sub-Saharan Africa is warranted. The countries of sub-Saharan Africa are largely poor and frequently they are poorly governed. In addition, they are too often very reliant on agriculture to sustain local populations, vulnerable to climate change, and, because of these limitations, badly in need of effective adaptation strategies to forestall mass starvation. Fortunately, many of the most highly effective adaptation strategies are being practiced in this region of Africa.

## Evergreen Agriculture

*Evergreen agriculture* is defined as an agricultural system in which trees (typically perennial) are integrated into the production of food crops (typically annuals).[26] This agricultural practice has had particular success in parts of Africa where there is a concurrence of undernutrition, drought, and increases in population. *Faidherbia albida* is the tree species at the center of this movement. This nitrogen-fixing acacia is indigenous to Africa. Its main (or tap) root penetrates deeply into the ground, making it quite resistant to drought. Other aspects of its annual cycle make it well suited to evergreen agricultural practices. Its dormant period coincides with the beginning of the rainy season. During dormancy, these trees lose their leaves and require little water. The fallen leaves provide organic matter that enrichs the soil even as the bare trees allow sunlight to penetrate to the sprouting crops planted beneath them, enhancing growth. The leaves reappear at the end of the wet season and shade the crops at this critical stage of crop growth. Combine these features with the acacia's nitrogen-fixing ability and you have a formula for success.

Other features of evergreen agriculture make it a highly desirable strategy, particularly in parts of Africa where per capita income is so low that farmers cannot afford to purchase fertilizer:[27]

- It maintains a cover of vegetation all year.
- Nitrogen fixation stimulates crop growth without inorganic fertilizer.
- Pests and weeds are suppressed.
- Water infiltration into the soil is enhanced.
- Soil structure is improved.
- The growth of the trees above and below ground captures carbon dioxide from the atmosphere.
- Biodiversity is enhanced.

Best of all, the production of food for the farmer and his family is improved, there is more fodder for animals, and more fuel for cooking. Frequently, evergreen agriculture–based farms produce enough food to sell some to others or for export, thereby improving the economic status of these low-income farm families. These evergreen practices are worthwhile now, but it is likely that they will prove to be even more useful in the future in protecting against climate change-induced problems.

Several regions of Africa already benefit from this practice. There has been a remarkable turnaround of the ecosystems and hence the fortunes of the people living in the Maradi and Zinder regions of Niger.[28] Niger lies in the middle of the Sahel region of sub-Saharan Africa between Libya and Algeria to the north, Nigeria to the south, and Chad and Mali to the east and west, respectively. Beginning in the 1960s, this region of Africa was thought to be in the midst of what might have become a state of irreversible decline. The desert was advancing, crops were failing, and firewood was scarce and becoming scarcer. Severe drought gripped the nation in the early 1970s. Livestock herds shrank dramatically, mortality rates climbed, and famine became widespread. The situation began to improve toward the end of that decade, when the knowledge of the indigenous people and nongovernmental organizations supplanted immediate postcolonial practices. Key among these was the rediscovery of the benefits of something as simple as proper pruning of what were thought to be shrubs. Pruning allowed these presumed shrubs to grow into trees, which provided shade to crops, which began to flourish. Altogether, over five million hectares (over 5.47 million acres; one hectare = 10,000 m$^2$, or 2.471 acres) were restored to productivity, in part by planting two hundred million trees that improved the livelihood of around 4.5 million people. Row crops such as millet, sorghum, peanuts, cassava, and others were planted between the trees and began to provide food and cash. Previously starving people began to export food to their southern neighbors. As an additional benefit, time spent on the task of foraging for firewood fell from three hours to thirty minutes per day. This change empowered women, those who had been tasked with collecting wood for cooking fires, thus enabling them to assume greater responsibilities in their homes and society.

Zambia is a successful variation on this theme. Maize is the country's primary crop.[29] Before the adoption of evergreen agricultural practices, yields were low, just over a ton per hectare (t/ha). Around 70 percent of farmers could not buy fertilizer, and about that same number failed to produce enough maize to sell at the market. Between 2002 and 2008, a third of the maize area under cultivation was abandoned before the harvest due to drought, falling soil fertility, and other factors. Here enters evergreen agriculture. Planting *Faidherbia albida* was combined with other improvements in agricultural practices fostered by the Zambian Conservation Farming Unit. Changes included the introduction of minimum tillage methods, cessation of burning crop residues from prior harvests, rotating crops, planting crops in precise locations to optimize fertilizer applications, and others. In 2008, at one site where *Faidherbia albida* trees had been planted the maize yield was 4.1 t/ha under the tree canopy, compared to 1.3 t/ha away from the canopy. In other regions, increases in yield of 280 percent were recorded.

Excellent results have also been reported in Malawi, where the economy is highly dependent on agriculture. In the pre-evergreen-agriculture era, more than half of farm households fell below subsistence levels, and large numbers of families required food aid. Governmental subsidies to provide fertilizer and the spread of agroforestry practices now have led to improvements. In one area of the country, maize yields under the *Faidherbia albida* trees were 100 to 400 percent higher than yields of maize not grown under the canopy of the trees. Malawi has a relatively long history of intermixing nitrogen-fixing trees with food crops, and trees other than *Faidherbia albida* also have been employed with success.

Now is the time to implement these strategies, before climate change results in further weakening of societies that are already fragile and increasing human suffering. Such steps will not solve all of the problems associated with climate change, but these are examples of some of the measures that can be enacted to adapt to and minimize the consequences of climate change.

## Rising Sea Level

As with agriculture, successful adaptation to rising sea level and the concomitant threat posed by storms and storm surges is highly dependent on local conditions. Successful coastline adaptation depends on local, state, and federal governance issues. Because of these complexities, detailed accounts are far beyond the scope of what can be covered in this book. Approaches to this issue that are particularly oriented around developing nations can be found in several sources, such as the United Nations Development Program's document "Adaptation Policy Frameworks for Climate Change: Developing Strategies, Policies, and Measures," and the United Nations Framework Convention on Climate Change's National Adaptation Programs of Action. However, this section offers a look at adaptation practices in two regions in the developed world that face similar challenges but have vastly different responses: South Florida and the Netherlands.

In chapter 1, I presented a general strategy for adaptation, in which the importance of political leadership, stakeholder involvement, the availability of funding, and other factors were discussed. In chapter 6, meanwhile, I presented three basic adaptation strategies: abandonment, in which adaptation measures either will not work or are deemed too expensive to implement; nourishment, in which natural barriers are augmented; and armoring, in which barriers to rising sea level and surges, such as sea walls, are constructed.

Dealing with the sea has been a serious issue in the Netherlands since the ninth century, when it is believed that the first dikes were constructed. According to a 2011 report, nine million Dutch people live in areas that are below sea level, and 70 percent of the country's GDP is generated in these areas.[30] In addition to protecting the population from the sea, other important aspects of the coastal zone of the Netherlands are related to industry, recreation, residences, supplies of drinking water, and other factors. Protecting these resources is a primary function of the Dutch government.

By modern standards, the early dikes were primitive. They consisted of earthen structures reinforced with logs and seaweed. Sediments tended to accumulate on the seaward side, making it possible to construct another dike in a series of seaward-marching structures. Skip ahead to the twentieth century, when in 1918 the Zuiderzee Act was passed and plans were launched to construct a huge dike to seal off an inlet of the North Sea known as the Zuiderzee from the rest of the North Sea. This effort was to be coupled with strategies to drain the newly reclaimed area. This project is known as the *Zuiderzee Works*. The dam is called the *Afsluitdijk*, and the sealed-off portion is the *Ijssulmeer*. I saw this dike myself in 1963, when a friend and I rode bicycles from one end to the other on a trip through Europe. It is an impressive structure.

The Dutch did not stop there. To ensure safety of the vulnerable shoreline, three more laws were passed: the Delta Act, the Flood Defense Act, and the Water Act.[31] Since 1990, the Dutch have relied heavily on nourishing the coastal zone by building up sandy areas, including dunes. Specific design criteria have been developed and implemented for the shape and size of these sandy shoreline defenses. Some of the sand used for this purpose is dredged up from the North Sea bed.

The Delta Works consists of a system of storm surge barriers and dams that was constructed in response to severe floods in 1953. The largest of these is the Eastern Scheldt storm surge barrier (in Dutch, *Oosterschelde-kering*). This barrier is nine kilometers long and has doors that can be opened and closed. It maintains the mix of sea and fresh water needed to support existing ecosystems while also providing protection from storms and storm surges. The construction of the Maeslant Barrier, shown in figure 10.3, was the final step in the Delta Works project. This movable barrier protects the harbor at Rotterdam. It consists of two semicircular gates that are hinged and that rotate to open or close the barrier. It cost €450 million to build and is said to be one of the largest movable structures on the earth. The barrier is controlled by computers linked to storm forecasts. When a storm surge greater than three meters is likely, a sequence of events is initiated that culminates in the closure of the barrier.

Sea level is a moving target, and the Dutch move along with it. They continue to seek input from stakeholders and experts in the field, as evidenced by surveys conducted to assist in the ranking of adaptation

**Figure 10.3**

Maeslant Barrier during a test closure. This storm surge barrier was built to protect the Rotterdam Harbor. It was completed in 1997. When monitoring systems predict a surge of more than three meters, computerized systems flood the dry docks that house each gate. The floating gates are moved to close the 360-meter-wide Rhine River waterway. Once in place, the gates are flooded and sink to prevent the surge from flooding the city and the port. *Source:* Koninklijk Nederlands Meteorologisch Instituut, http://bit.ly/19JEIiJ, accessed April 1, 2015. Not copyrighted; original in color.

options. The qualitative portion of a 2009 assessment focused on prioritization and ranking of various options. Results included a need to integrate water and nature management, integrate coastal zone management, and provide more space for water in rivers and regional water systems. The quantitative evaluation identified costs and benefits of the various options for adaptation.[32]

The Dutch take evaluating the risks posed by rising sea level seriously and have spent and are prepared to spend additional large sums to provide protection for the citizens of their nation. They are determined to prevent potentially catastrophic social and economic upheavals caused by climate change.

This is in stark contrast to South Florida and the Florida Keys. According to an article in the *Miami Herald* published in March 2015, state officials work actively to discourage any reference to climate change.[33] This effort is apparently led by Governor Rick Scott, who took office in 2011. The article provides detailed accounts, all transmitted verbally with no written policy, that discourage multiple state agencies from making references to climate change and related topics.

The geology of South Florida places limits on adapting to rising sea level. The coral limestone that is found throughout the region is porous; that is, water moves through it quickly. This is an advantage during the heavy rains that are common in the region. Within a matter of minutes, rainfall of up to a foot over a few hours soaks into the ground and disappears—a good thing. However, building sea walls or physical barriers as armor against rising sea level will fail because the ocean's water will seep under the barrier through pores in the rock. Infiltration of sea water into fresh water supplies is already a problem.

In many areas, such as South Miami Beach and Miami Beach, large amounts of sand are dredged up from the ocean floor to nourish the beaches that draw the tourists. For many areas in the Florida Keys, placing buildings on stilts is the only practical solution.

Local governments serving South Florida and the Florida Keys are not constrained by gag orders, but they lag far behind the Dutch in planning for increases in the sea level and storm surges. The Florida Keys are a curved group of around 1,700 sandy islands perched on coral reefs and rocks that extend around 220 miles from just south of Miami to Key West. The average elevation of the Florida Keys is 1.5 meters above sea level, in a part of the world that is highly exposed to hurricanes. In spite of their vulnerability, the islands have a hugely popular, multibillion dollar economy built largely on tourism. In 2011, two professors from Florida International University in Miami, along with a Massachusetts colleague, published the results of a survey conducted among stakeholders who serve the Florida Keys.[34] Their anonymous survey included federal, state, and local officials, along with representatives from nongovernmental organizations, such as the Audubon Association. A total of 845 potential participants were contacted; 26.6 percent completed the questionnaire. The low response rate is problematic. Of the respondents, 9.6 percent were federal officials and 17.6 percent were state officials. The authors of the 2011 report found a "deep concern among federal, state, and local decision makers and experts." The most discouraging result of the survey was that although 85.7 percent of the respondents support preparing now for the events that are the most likely to occur as the result of climate change, only 5 percent indicated that their agencies or organizations had a climate change adaptation/action plan.

A three-county coalition in Southeast Florida (Miami-Dade, Broward, and Monroe Counties, the latter home to the Florida Keys) reported on their joint efforts to begin to address climate change with a report titled "A Region Responds to a Changing Climate" in October 2012.[35] Ironically the report's cover art features a woman wearing a chic dress riding a motor scooter on a flooded street in Miami-Dade County as though she was on a vacation in Rome and not facing an environmental disaster. The goal of the coalition was to "unite, organize and assess [its] region through the lens of climate change in setting the stage for action." The Compact, as members refer to the group, calls for "urging Congress to pass legislation that recognizes the unique vulnerabilities of Southeast Florida to climate change impacts, especially sea level rise," along with other objectives that depend largely on federal action. Of the thirteen public policy objectives, three began with "Urge Congress," another with "Encourage federal support for research," and four with either "Advocate" or "Support and advocate."

The contrasts are stark. The people of the Netherlands have been building defenses against the sea for over one thousand years; the people of South Florida are giving serious thought to the process, or at least some are, but so far they have done little. Whether Floridians meet the challenge is yet to be determined.

## Clean Coal

The terms *clean coal* and *war on coal* are often couched in other terms, such as *Obama's war on coal* or *Obama's war on jobs*, which emphasize the political split in the United States as opposed to a sincere effort to deal with coal-derived pollution.

Most of the time, *clean coal* refers to a series of processes known collectively as *carbon capture and storage* (CCS). However, some use *clean coal* to describe other aspects of coal use. Some cleaning typically occurs at the mine itself, where coal is washed. When it emerges from the mine, the coal is contaminated by unwanted waste. This waste includes dirt, rock, sulfur, and other noncoal debris. The coal is washed by utilizing differences in the physical properties of coal and waste to divert the waste to a repository, which makes it more economical to ship the final product. The downside of this washing process is that it creates yet another waste stream that must be dealt with. The residents of Buffalo Creek Hollow, West Virginia, learned about this the hard way when a series of dams designed to impound the coal-washing-waste slurry failed after heavy rains.[36] The resulting flood was responsible for 125 deaths and 1,100 injuries and rendered four thousand people homeless. It was one of the worst flood disasters in US history.

There is substantial corporate interest in so-called clean coal. One company, Clean Coal Technologies, with corporate headquarters in New York City, has a website that features photographs of children, their mothers, and dogs. The company's technology is aimed at the overseas markets for coal. It has developed a process that removes water and hydrocarbons from coal, making it cleaner, lighter, and cheaper to ship. The company also leads people to believe that its product is less likely to undergo spontaneous combustion, which is a significant problem for low-grade coals. I saw many coal fires at a huge lignite mine in Kosovo, and all were the result of spontaneous combustion of poor quality coal, virtually the only natural resource in that impoverished nation.

Now the term *clean coal* is used almost exclusively to refer to efforts to prevent carbon dioxide formed by burning coal from entering the atmosphere. With the exception of a few demonstration projects, CCS technology exists only in the realm of the future. We will return to the topic of CCS in the section of this chapter describing climate interventions.

### The Clean Power Plan

On June 2, 2014, the EPA published what it called a commonsense rule designed to limit carbon dioxide emissions from existing power plants.[37] In a decision supported by the US Supreme Court, the EPA claimed that its authority to regulate carbon dioxide emissions is derived from the Clean Air Act. Those who believe that this gas drives climate change and poses an enormous threat to all hailed the plan as a long overdue step, whereas climate change deniers lined up in opposition to what they viewed as yet another step taken by the government to control our lives.

The plan, if implemented, is designed to reduce carbon dioxide emissions from existing power plants by 32 percent by the year 2030, compared to the 2005 baseline. Like most EPA rules, it is filled with specific numbers, but essentially the proposed rule will limit emissions from coal-fired plants to no more than those expected from power plants fueled by natural gas.

Each state will be required to evaluate its strengths and weaknesses and design a state-specific approach to fulfilling its obligation. In general, the plan encourages states to design plans that convert coal plants to natural gas; increase the amount of electricity generated from renewable sources, such as wind, water, and solar; and improve the efficiency with which power is used. The EPA notes that as a nation we are well on our way toward meeting the 2030 goal.

In its analysis of the impact of the plan, the EPA claims that when implemented, compliance will cost between $7.3 billion and $8.8 billion per year but will save between $55 billion and $93 billion annually. The savings are largely in the realm of better health as a result of lower emissions of criteria pollutants and climate change prevention. Mean estimates for emission reductions are around 450,000 tons of sulfur dioxide, just under 420,000 tons of nitrogen dioxide, and 55,000 tons of small particles. These translate into between 2,700 and 6,600 fewer deaths, from 140,000 to 150,000 fewer attacks of asthma in children, between 340 and 3,300 fewer heart attacks, from 2,700 to 2,800 fewer hospital admissions, and between 470,000 and 490,000 fewer missed days at school and work.

The Clean Power Plan will have two benefits. First, it will improve the health of individual Americans. Second, it will begin to slow the pace of climate change. It will not prevent all of the effects of climate change, however; at best, it may get us off the trajectory predicted by the IPCC business-as-usual scenario. If this plan is successful and if other nations follow through with effective measures to reduce greenhouse gas emissions in order to combat climate change, it is possible that the political leadership needed to combat climate change will emerge along with support by stakeholders. The 2015 UN Conference on Climate Change could be an important turning point in the efforts to mitigate climate change.

In February 2016, the Supreme Court unexpectedly blocked implementation of the Clean Power Plan in a 5 to 4 vote. Within days, Justice Antonin Scalia died and Senate Republicans appeared to be poised to block any appointment by President Obama. Thus the future of the plan may be in doubt. If the Court blocks the plan, the climate change accord reached by the 2015 UN Conference could be thrown into disarray.

**Powering New York with Wind, Water, and Solar Energy**

In 2013, professors from Stanford and Cornell published a detailed roadmap showing how virtually all of New York's energy needs could be met using readily available wind, water, and solar (WWS) off-the-shelf technology.[38] Electricity would become the dominant source of power. It would be generated by onshore and offshore wind turbines (just over sixteen thousand five-megawatt [MW] units), solar-photovoltaic (PV) plants (just over 820, generating 50 MW each), residential rooftop systems (five million at five kilowatts [kW] each), and PV systems placed on commercial and governmental buildings (around five hundred thousand 100 kW systems). The plan also envisions around eight hundred concentrated solar systems generating 38,700 MW. These sources would be supplemented by geothermal systems to supply around 5 percent of the state's energy demands, tidal systems to supply 1 percent of the need, and hydroelectric systems to supply 5.5 percent of the need (of which almost all exist today). The transportation sector, with the exception of air traffic, would be powered by hydrogen fuel cell power sources equipped with regenerative energy capture systems, such as those already in use in hybrid automobiles, buses, locomotives, and trucks. Residential, commercial, and governmental buildings would be heated by electrical resistive units. Batteries, some of which would be in vehicles and be able to power the grid at night; storing heat and cold in designed sinks; and splitting water molecules to form hydrogen and oxygen would all power the grid at night and when there is no wind.

Gains in efficiency due to some grid modernization and the relative efficiency of electricity over fossil fuels would result in a projected 37 percent overall power saving. Some improvements in efficiency would be achieved by the replacement or de novo installation of efficient appliances, the use of LEDs for lighting, and similar steps designed to decrease the use of electricity.

The authors of the New York proposal provide an accounting of costs associated with the migration to sustainable WWS power. True, there would be capital and maintenance costs associated with the installation of turbines and so on. However, once installed, there would be large savings because the energy to power the WWS generators is free. The cost of electricity between 2020 and 2030 under the proposed plan is estimated to be between $0.04 and $0.11 per kilowatt-hour (kWh), including transmission and distribution. This is substantially less than the estimates of $0.178 and $0.207 per kWh for electricity generated by burning fossil fuels. These cost estimates include not only the cost of electricity itself but also the so-called externalities, such as health costs associated with the mortality and morbidity associated with air pollution from fossil fuel combustion. These costs are real, but they are rarely included in reports of the "real" costs of power.

New York produces only small amounts of fossil fuels, so few jobs would be lost in this sector of the economy. This job loss would be more than compensated for by the large number of new jobs that would be created to construct and maintain the new electricity-generating units. Overall, the plan is a net job creator.

These sources of electrical power do not produce hazardous air pollutants such as oxides of sulfur and nitrogen, small particles, carbon monoxide, and others that are harmful to health. Thus, pollution-related morbidity and mortality would fall, along with corresponding health care costs, and the cost of economic opportunities lost due to pollution-related job impacts. The median estimate for this cost savings is $33 billion per year, which is around 3 percent of the gross domestic product for the state. By the year 2050, climate change costs to the nation would fall by an estimated $3.3 billion per year.

When something sounds too good to be true, it probably is. In this case, it is almost too much to hope for; political will among our leaders, who have become followers, is the missing ingredient.

## Solutions beyond Medicine

Thus far, I have largely presented relatively conventional measures that fit
under the umbrella of *health*, as defined broadly by the World Health
Organization. In the final portion of this chapter, I will turn to more
extreme measures that are advocated by some as reasonable, technologi-
cally based strategies to combat climate change. This calls to mind one of
the aphorisms of Hippocrates: "For extreme diseases, extreme methods
of cure … are most suitable."

No matter what decisions we will make as a civilization on our disor-
ganized and occasionally contradictory paths toward the future, we must
curtail the emissions of greenhouse gases. Failure to do so will almost
certainly lead to a climate legacy that none of us want to leave for our
children, grandchildren, and those who will follow.

On one hand, there are proponents of rapid movement toward sustainable energy sources that use available, off-the-shelf technology; on the
other, there are proponents of solutions that require substantial amounts
of research and development. The first position is exemplified by plan
for New York, discussed earlier in this chapter. The second relies on technologies that are under development, the focus of the remainder of this
chapter.

## Climate Interventions: Managing Carbon Dioxide Emissions with Advanced Technology

In the spring of 2015, the National Academy of Sciences (NAS) issued
the findings of its Committee on Geoengineering Climate. The committee conducted a critical evaluation of the science, risks, and potential
benefits of selected strategies designed to cope with climate change.
The NAS prefers the term *climate intervention* to one that may be more
familiar, *geoengineering*. The rationale is that *engineering* implies a
higher level of precision and certainty than warranted by the evidence.
The committee presented its findings in two reports, one based on solar
radiation management and the second on carbon dioxide removal technologies.[39] The studies were supported by NAS, NASA, NOAA, and,
somewhat surprisingly, by the "US Intelligence Community." The reports
can be downloaded at no cost from NAS's web site (http://www
.nationalacademies.org).

**Carbon Capture and Reliable Sequestration**

The NAS report on managing carbon adds the word *reliable* to the standard terminology, *carbon capture and storage*. This emphasis on the reliability of the storage is entirely appropriate. It does the world little good if the carbon dioxide that is stored escapes into the atmosphere. I also find the use of the word *carbon* as shorthand for *carbon dioxide* to be somewhat confusing and even misleading. Perhaps a linguistics consultant decided that the word *carbon* is less likely to trigger opposition than the full name of the greenhouse gas that we must deal with.

Some approaches to managing carbon dioxide require little, if any, technology. They are based on improving land-management practices. As shown in figure 2.3, which illustrates the simplified carbon budget for the earth, between 1.7 and 2.6 trillion tons of carbon (not carbon dioxide) are sequestered in soils, and between 495 and 715 billion tons are trapped by vegetation—a good thing and an explanation for the annual dip in atmospheric carbon dioxide shown in the Keeling Curve (see chapter 2). This movement of carbon from the atmosphere to the land amounts to around three billion tons per year. Land-management strategies could be redesigned to minimize loss of carbon from the soil and vegetation and maximize the movement from the atmosphere to the earth stores.

The NAS report indicates that deforestation released approximately three gigatons of carbon dioxide annually between 2002 and 2011. Most of this deforestation occurs in the tropics as land is cleared and trees are burned to make way for crops and grazing. Deforestation accounts for around 10 percent of all anthropogenic greenhouse gas emissions. An examination of the seasonal dips in the atmospheric carbon dioxide concentration shown in the Keeling Curve (see figure 2.4) demonstrates the potential for removal of carbon dioxide from the atmosphere by forests. Each spring and early summer, the forests in the northern hemisphere begin to grow new leaves and branches. This is reflected in the transient dip in the atmospheric carbon dioxide concentration seen at that time of the year.

The growth of cover crops on land that is not producing food will also remove carbon dioxide from the atmosphere. Plowing these crops into the soil will sequester additional amounts of carbon dioxide in the soil.

Another way to capture carbon dioxide is to convert it into limestone, which contains large amounts of calcium carbonate. When this mineral is exposed to carbon dioxide, it forms calcium and bicarbonate ions, as shown in the following equation:

$CO_2 + CaCO_3 + H_2O$ react to form $Ca^{++} + 2\ HCO_3^-$

When dissolved, bicarbonate ions enter seawater, which becomes more alkaline. This could be a good thing, because carbon dioxide dissolving in our oceans has made them more acidic, thereby threatening marine eco-systems. Eventually, the bicarbonate ions are trapped in the shells of marine animals as carbonate, thus storing the carbon. Similar reactions involving carbon dioxide and carbonate-containing minerals may prevent at least some carbon dioxide from traveling to the surface after a deep well injection. This process is envisioned by some as a means to capture and sequester carbon dioxide formed by burning fossil fuels, as discussed ahead.

Planktonic algae and other plants that live in the ocean remove carbon dioxide from the surface layers of the ocean and convert it to sugars and other organic molecules by photosynthesis. This trapped carbon either will be eaten by marine animals or will settle to the bottom of the ocean. This natural process has attracted the attention of some who have pro-posed stimulating these reactions by fertilizing the ocean with iron. Both controlled and poorly controlled experiments have been performed to test this strategy; one of the first of these is referred to as the IronEx II experi-ment and was conducted in an equatorial region of the Pacific Ocean.[40] The success of the trial is reflected in the title of the paper describing it: "A Massive Phytoplankton Bloom Induced by an Ecosystem-Scale Iron Fertilization Experiment in the Equatorial Pacific Ocean." Other so-called experiments appear to have been conducted by rogue individuals. One such study was alleged to have been conducted by an American business-man who claimed to have dumped one hundred tons of iron-rich, dirt-like material into the ocean in 2012.[41] Herein lies one of the problems associ-ated with techniques of this type: so-called experiments can be performed by virtually anyone, without scientific, technical, or ethical oversight or international cooperation.

The most conventional process for the removal of carbon dioxide pro-duced by burning fossil fuels is generally referred to as carbon capture and sequestration or storage. In broad terms, a fuel is burned and the resulting carbon dioxide is captured, liquefied, transported to a secure repository, and sequestered for periods measured on a geological scale. The fuel source may be biomass or fossil in origin, or some combination of the two.

A recent analysis of the use of biomass reveals some of the limits of this strategy.[42] Large amounts of land are required to grow, for example, switchgrass, which needs to be fertilized with both nitrogen- and phosphorous-containing fertilizers. In addition, water is required, a problem for areas already stricken by drought or areas where droughts may develop. Finally, the process is not likely to be very energy efficient. The example provided in the study predicts an overall energy efficiency of just over 47 percent. The greatest losses occur during the processing needed at the electrical generating unit to prepare the grass for burning (62 percent efficient) and the actual capture process (89 percent efficient).

The more typical consideration, and the one promoted most heavily by the coal industry, involves burning coal.[43] There are several different processes in the combustion step, ranging from more or less straight combustion to more sophisticated strategies that involve gasification. Some use ordinary air, whereas others use highly enriched oxygen. The goal of the latter strategy is to produce a post-combustion gas that is highly enriched in carbon dioxide.

The use of the term *capture* is misleading and may be a source of some confusion. For the purposes of the carbon (dioxide) capture and storage (CCS) literature, *capture* means producing a flue gas that has a very high percentage of carbon dioxide—the higher, the better. When the police capture a suspect, one envisions a set of circumstances that preclude escape—not so with CCS. The carbon dioxide will easily enter the atmosphere unless energy-intense processes are used to keep this from happening. The next all-important step involves taking the captured carbon dioxide and compressing it so that it becomes a liquid.

The liquefied carbon dioxide must then be transported to a site where it can be sequestered from the atmosphere. Various studies have shown that pipelines are likely to be the cheapest way to transport liquid carbon dioxide. There are dangers inherent in transporting any product by pipeline. In the case of carbon dioxide, the greatest risks are associated with rupture and leakage. Although carbon dioxide does not burn—in fact, it is used widely in fire extinguishers—it is toxic and can cause asphyxiation and death at concentrations that are around 17 percent by volume.[44] Since carbon dioxide is heavier than air, any gas that escapes from pipelines is likely to settle in low-lying areas or basements, where it may be undetected by anyone who enters the area. A disaster near Lake Nyos in Cameroon illustrates the dangers this fact poses.[45] Lake Nyos covers volcanic magma and carbon dioxide. On October 2, 1986, there was a natural release of carbon dioxide from under the lake that killed around 1,700 individuals as the gas flowed downhill from the lake and into villages.

The final step in the process, and perhaps the most difficult, is permanent storage in a manner that keeps the $CO_2$ separated from the atmosphere—forever. The strategy proposed most frequently relies on injection into deep wells that are specially designed and constructed for this purpose. Supporters of this choice note that this process has been used widely to enhance the recovery of oil from wells that were running dry. Although earthquakes attributed to injection of wastewater from hydraulic fracturing into wells have been reported in Ohio, this problem has been minimized by the proponents of injection—even though the practice has been stopped in that state. It is likely that a report issued jointly by the Oklahoma and US Geological Surveys in the spring of 2015 will alter the perception of earthquake safety substantially.[46] Between 1978 and 1999, Oklahoma had an average of about 1.6 earthquakes that measured 3.0 or higher on the Richter scale per year. The Richter scale is logarithmic; every unit increase represents a tenfold increase in the amount of energy released. Beginning in 2009, about the time that deep well injection of fracking waste began in earnest, the number of earthquakes began to rise. There were twenty that year. There were 109 in 2013 and 584 in 2014, of which nineteen exceeded 4.0 on the Richter scale. The USGS estimates that by the end of 2015 there will be 941 earthquakes of magnitude 3.0 or greater if the current pace continues without any change. These data are shown in figure 10.4.

In its report, the NAS states clearly that there is no substitute for reducing carbon dioxide emissions. Without this critical commitment, all other strategies have serious drawbacks. For the plans designed to remove carbon dioxide from the atmosphere, they make the additional points:

- Removing carbon dioxide addresses the most important cause of climate change—high greenhouse gas concentrations.

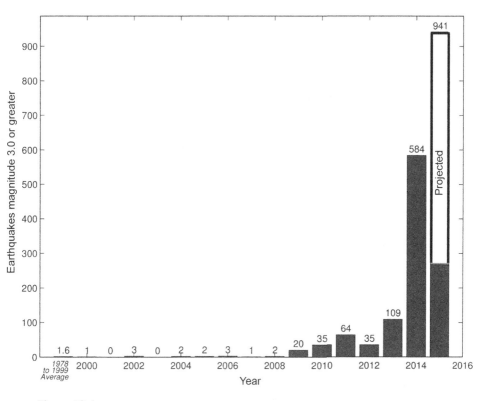

**Figure 10.4**

The earthquake history of Oklahoma. The USGS estimates that there will be 941 earthquakes with a magnitude of 3.0 or greater for the year 2015 if the frequency continues to accelerate at the rate observed in the spring of that year. In the interval between 1978 and 1999, the state averaged 1.6 earthquakes of that magnitude annually. Modified from a color graph published by the United States Geological Survey. United States Geological Survey, "Oklahoma Earthquake Information," last updated April 18, 2014, http://earthquake.usgs.gov/earthquakes/states/?region=Oklahoma, accessed December 29, 2015.

- Unlike other proposed solutions, these plans do not pose risks on a global scale (earthquakes are local, not global, in their nature).
- The plans have a high cost and are likely to be judged solely on cost.
- The effects will be modest at best because of the long-lived nature of atmospheric carbon dioxide (a lifetime of thirty to thirty-five thousand years). Large-scale implementation by major carbon dioxide emitters is a prerequisite to success.

- These plans do not require new international agreements before implementation.
- Incremental implementation is possible.
- Unlike some interventions designed to mitigate climate change, abrupt termination would have only a small effect (see below).

Coal companies are among the chief proponents of CCS. Is this a legitimate position that is designed to mitigate climate change? Or, as claimed by the cynics, is it merely a device to promote mining and sell more coal at a time when natural gas is replacing coal in many power plants? The answer might be found by following the money. The money issue was detailed in an August 2015 report in the *New York Times*, titled "King Coal, Long Besieged, Is Deposed by Market."[47] At least four large coal producers have declared bankruptcy, and the stock prices of others have fallen drastically. Finally, we should ask ourselves who is paying for the public relations campaign that promotes CCS.

A final question remains. Even if CCS were to operate at the highest projected level of efficiency, would it prevent enough carbon dioxide from entering the atmosphere to make a difference?

### Albedo Modification

*Albedo* is a term that describes the degree to which solar energy is reflected back into space by the earth. High albedo regions reflect lots of energy, whereas low albedo regions reflect little or none. Measures that increase the earth's albedo lead to cooling, and the reverse also is true. An example of this effect is the fact that the loss of the highly reflective snow and ice around the Arctic has led to a decrease in the earth's albedo that favors warming. Increases in the earth's albedo after huge volcanic eruptions that inject sulfates and dust into the atmosphere have produced transient cooling. This effect was confirmed by direct measurements made after the June 1991 eruption of Mount Pinetubo in the Philippines.[48] Many believe that the 1815 eruption of Mount Tambora led to what has been referred to as *the year without a summer*. This cataclysmic event is the centerpiece of Gillen D'Arcy Wood's new book, *Tambora: The Eruption That Changed the World*.

Volcanic eruptions of the type exemplified by the Pinetubo and Tambora events have complex effects on the climate related to the injection of massive amounts of sulfates and dust into the stratosphere. This injection reflects solar energy back into space. However, important provisions of the regulations promulgated under the authority of the Clean Air Act are designed to reduce sulfates and particles in the air. From a technical perspective, the products of the eruptions alter the atmospheric aerosol. *Aerosols* are fine suspensions of solid particles and liquid droplets in a gas. The portion of the NAS report that concentrates on albedo modification focuses on strategies designed to modify the stratospheric aerosol or, alternatively, to change the albedo of the clouds over oceans.

Although volcanic eruptions are known to have effects on climate if they are large enough, incorporating this strategy into a realistic attempt to cool the earth cannot be justified on the basis of the data at hand. This is exemplified by a quote from the NAS Climate Interventions report: "No well-documented field experiments involving controlled emissions of stratospheric aerosols have yet been conducted."[49] Thus, this is an area where an enormous amount of research that has yet to be undertaken must be performed before any evidence-based decisions can be made.

There also seems to be little to be gained by injecting sulfur dioxide into the atmosphere and a lot to lose. Although this action would lead to cooling, it would move the earth in a direction that is exactly the opposite of the intent of the Clean Air Act. Sulfur dioxide is one of the criteria pollutants listed by the EPA. It is a highly reactive chemical that irritates the lungs and other tissues. Sulfur dioxide also reacts with other elements in the atmospheric aerosol to form fine particles. These too are criteria pollutants. Particulate matter and sulfur dioxide have large detrimental effects on health. So, one might think of enacting sulfate injections as jumping out of the frying pan and into the fire.

The case is only marginally better when considering altering the clouds over the oceans to reflect sunlight. Clouds at low altitudes that cover large portions of the earth's oceans scatter sunlight back into space. These clouds reflect energy away from the planet that would have been absorbed by the darker water beneath the cloud layer, which forms the basis for considering modifications to the cloud layer as an earth-cooling measure.

Some proof-of-concept data are the result of several experiments reviewed in the NAS report. Under very specific and limited conditions, diesel exhaust particles emitted by ships at sea may serve as a nidus for the condensation of water vapor to form clouds. These clouds are somewhat analogous to the contrails produced by high-altitude aircraft. However, these ships consume large amounts of fuel—around one hundred thousand gallons per day. This would almost surely create an unacceptably large carbon dioxide burden. There is no such thing as a free lunch.

Other, more fanciful strategies include blocking solar radiation with mirrors or other devices launched by rockets, manipulating the albedo of the earth's surface, and manipulating cirrus clouds. This cloud layer is composed of ice crystals that contribute to greenhouse warming, the dominant effect, and reflect solar energy, preventing warming. By decreasing their opacity, distribution, and lifetime, it might be possible to affect the climate in the short term.

Aside from the all-important fact that little is known about these potential technological fixes, the NAS report lists other potential complicating factors:

- These measures do nothing to address the underlying cause of climate change: uncontrolled emissions of greenhouse gases.
- They pose significant risks that are global, novel, and poorly understood.
- There are no international bodies that could exercise oversight.
- Unilateral, even rogue implementation creates the potential for unforeseeable threats.
- Not every place on earth would be affected evenly. Thus, unilateral actions could benefit some at the expense of others. Could this create the potential for climate warfare?
- Abrupt cessation of these "fixes" would lead to rapid warming, because greenhouse gases would continue to accumulate during periods of active interventions.
- Because carbon dioxide would continue to accumulate in the atmosphere, other effects of the gas would continue to increase, including ocean acidification, effects on agriculture, and so on.

It is likely that these strategies would be much less expensive than methods that capture and sequester carbon dioxide. They could also act rapidly after a perceived climate emergency. Some methods are capable of acting in time spans measured in years rather than decades or centuries.

At this point in time, climate engineering solutions appear to be a mixed bag of potential benefits and losses.

## Miscellaneous Considerations

In their plan to convert New York's energy sources to wind, water, and solar, Jacobson and his colleagues envisioned storing electricity in batteries—presumably batteries in hybrid vehicles or similar battery arrays.[50] Owners of these vehicles would plug them into stations that would utilize this stored energy when needed—notably, when the sun was not shining or the wind was not blowing. Just a short time after the publication of their report, additional, and possibly more practical, strategies for battery storage of electricity have begun to emerge. At least two large firms are developing batteries for this purpose. Perhaps the most prominent and highly publicized of these batteries is being manufactured by Elon Musk's firm that makes lithium ion batteries for its Tesla automobiles.[51] Musk received substantial news coverage in February and April 2015 when he announced that a version of this battery, being produced in a $5 billion factory, would be marketed for home use. Earlier news coverage suggested that the factory would be able to produce five gigawatts of battery storage capacity by 2020. UniEnergy Technologies also has plans to market liquid flow batteries that are capable of storing a megawatt of power that can be discharged over three to four hours.[52] This amount of electricity would be capable of powering around five hundred typical homes. These batteries are about the size of the containers used for transoceanic shipping. From the company's website, it looks like they are designed to be moved easily, perhaps a huge advantage during an emergency such as a hurricane, flood, or other disaster. This is clearly a rapidly emerging field.

Energy can also be stored in the form of hydrogen, the most abundant element in the universe. A group of Chinese investigators appears to have made significant progress toward the goal of splitting water molecules into hydrogen and oxygen.[53] In a paper with the enticing but daunting title "Metal-Free Efficient Photocatalyst for Stable Visible Water Splitting via a Two-Electron Pathway," this group opens the door to producing large amounts of hydrogen without the need for any exotic metal catalysts. The group used carbon to make *nanodots*, defined as quasi-spherical carbon particles with a diameter of less than ten nanometers.

The researchers were able to achieve an efficiency of 2 percent, which, with further refinements, should make it possible to produce hydrogen gas on a commercial scale in a cost-effective manner, using sunlight as the energy source. The hydrogen could be stored and used at a later time in fuel cells or some other energy-producing device.

## The Need for Research and Education

In an April 2015 op-ed in the *New York Times*, former Republican Speaker of the House Newt Gingrich called for doubling the NIH budget.[54] He cited the appropriate data concerning the impacts of diseases on the American public, concluding with the statement that supporting research to facilitate medical progress was an appropriate role for the federal government. We should not stop with the NIH; this plea should extend to all federal and nonfederal research programs. President Obama used different words in his January 25, 2011, State of the Union speech when he called on us all to realize the huge American potential to "out-innovate, out-educate, and out-build" the rest of the world.[55]

## The Trajectory toward the Future

Experience has shown clearly what must be done to minimize the effects of climate change. Although there is no substitute for reducing greenhouse gas emissions, vigorous adaptive measures are also needed. We must be better prepared to cope with more heat waves. The public health infrastructure must be strengthened worldwide to cope with the threats posed by infectious diseases. Agricultural methods must be improved so that we can feed everyone. Solutions to the multiple dilemmas posed by rising sea level must be developed and acted on; we need to be more like the Dutch and less like Floridians. The root causes of violence require attention.

A number of targets have been set in order to minimize climate change. The group 350.org and its website 350.org were established with the goal of keeping the atmospheric carbon dioxide concentration below 350 ppm. That target was passed long ago and was probably never realistic. The World Bank called for keeping the global temperature increase below 4°C.[56] A lower target that seeks to limit the temperature increase to 2°C emerged from the Copenhagen conference.[57] Although that meeting failed to yield an agreed-upon enforceable goal, it did give the 2°C goal additional visibility. A recent report suggests that even this goal, which seems elusive, may not be sufficiently low to prevent a so-called tipping point, defined by the IPCC Fifth Assessment Report as "a large-scale change in the climate system that takes place over a few decades or less, persists (or is anticipated to persist) for at least a few decades and causes substantial disruptions in human and natural systems."[58] A report published in October 2015 identified thirty-seven of these events affecting oceans, sea ice, snow cover, permafrost, and terrestrial ecosystems. Of these, eighteen were anticipated to occur at or below the 2°C target. The earth's climate system may not be as resilient as we hoped.

We must join our partners across the world to accept and meet the challenge of climate change. There are huge barriers that must be overcome—and quickly. We lack the worldwide political leadership and the institutional organizations that are needed to identify and execute climate change policies. More stakeholder involvement is needed. At every stage, more research, development, and funding are needed. We must succeed. The cost of failure is too high to bear.

# Notes

## Preface and Acknowledgments

1. http://w2.vatican.va/content/francesco/en/encyclicals.index.html#encyclicals.
2. N. Watts, W. N. Adger, P. Agnolucci, et al., "Health and Climate Change: Policy Responses to Protect Public Health," *The Lancet* 386, no. 10006 (November 7–13): 1861–1914, http://dx.doi.org/10.1016/S0140-6736(15)60854-6.

## 1 Introduction

1. D. M. R. Rumsfeld, transcript of a Department of Defense news briefing of Secretary Rumsfeld and General Myers, February 12, 2002.
2. J. N. Pauli, J. E. Mendoza, S. A. Steffan, et al., "A Syndrome of Mutualism Reinforces the Lifestyle of a Sloth," *Proceedings of the Royal Society B: Biological Sciences* 281, no. 1778 (2014): 20133006.
3. N. Watts, W. N. Adger, P. Agnolucci, et al., "Health and Climate Change: Policy Responses to Protect Public Health," *The Lancet* 386, no. 10006 (2015): 1861–1914, http://dx.doi.org/10.1016/S0140-6736(15)60854-6.
4. C. Corvalán, S. Hales, A. J. McMichael, et al., *Ecosystems and Human Well-Being, Health Synthesis: A Report of the Millennium Ecosystem Assessment* (Geneva: World Health Organization, 2005).
5. Department of Economic and Social Affairs of the United Nations Secretariat, *Millennium Development Goals Report 2015* (2015); United Nations, *The Millennium Development Goals Report 2013* (New York: United Nations, 2013).
6. Ban Ki-moon, *The Road to Dignity by 2030: Ending Poverty, Transforming All Lives, and Protecting the Planet—Synthesis Report of the Secretary-General on the Post-2015 Agenda* (New York: United Nations, 2015).
7. R. Lozano, M. Naghavi, K. Foreman, et al., "Global and Regional Mortality from 235 Causes of Death for 20 Age Groups in 1990 and 2010: A Systematic Analysis for the Global Burden of Disease Study 2010," *The Lancet* 380, no. 9859 (2012): 2095–2128.

8. S. S. Lim, T. Vos, A. D. Flaxman, et al., "A Comparative Risk Assessment of Burden of Disease and Injury Attributable to 67 Risk Factors and Risk Factor Clusters in 21 Regions, 1990–2010: A Systematic Analysis for the Global Burden of Disease Study 2010," *The Lancet* 380, no. 9859 (2012): 2224–2260.

9. R. D. Brook, S. Rajagopalan, C. A. Pope III, et al., "Particulate Matter Air Pollution and Cardiovascular Disease: An Update to the Scientific Statement from the American Heart Association," *Circulation* 121, no. 21 (2010): 2331–2378; A. H. Lockwood, *The Silent Epidemic: Coal and the Hidden Threat to Health* (Cambridge, MA: MIT Press, 2012).

10. B. D. James, S. E. Leurgans, L. E. Hebert, et al., "Contribution of Alzheimer Disease to Mortality in the United States," *Neurology* 82, no. 12 (2014): 1045–1050.

11. D. Campbell-Lendrum, D. Chadee, Y. Honda, et al., "Human Health: Impacts, Adaptation, and Co-Benefits," in *Climate Change 2014: Impacts, Adaptation, and Vulnerability; Part A: Global and Sectoral Aspects; Contribution of Working Group II to the Fifth Assessment Report of the Intergovernmental Panel on Climate Change*, ed. C. B. Field, V. R. Barros, D. J. Dokken, et al., 709–754 (New York: Cambridge University Press, 2014)

12. IPCC, "Summary for Policymakers," in *Climate Change 2014: Impacts, Adaptation, and Vulnerability; Part A: Global and Sectoral Aspects; Contribution of Working Group II to the Fifth Assessment Report of the Intergovernmental Panel on Climate Change*, ed. C. B. Field, V. R. Barros, D. J. Dokken, et al. (New York: Cambridge University Press, 2014), 1–32.

13. J. B. Smith, J. M. Vogel, and J. E. Cromwell III, "An Architecture for Government Action on Adaptation to Climate Change: An Editorial Comment," *Climatic Change* 95, nos. 1–2 (2009): 53–61.

14. Ibid.

## 2   The Scientific Evidence for Climate Change

1. P. N. Pearson and M. R. Palmer, "Atmospheric Carbon Dioxide Concentrations over the Past 60 Million Years," *Nature* 406, no. 6797 (2000): 695–699.

2. IPCC, "Summary for Policymakers," in *Climate Change 2013: The Physical Science Basis; Contribution of Working Group I to the Fifth Assessment Report of the Intergovernmental Panel on Climate Change*, ed. T. F. Stocker, D. Qin, G.-K. Plattner, et al. (New York: Cambridge University Press, 2014), 4.

3. Ibid.

4. D. L. Royer, R. A. Berner, I. P. Montañez, et al., "$CO_2$ as a Primary Driver of Phanerozoic Limate," *GSA Today* 14, no. 3 (2004): 4–10.

5. IPCC, "Summary for Policymakers," 17.

6. UNEP 2012, *The Emissions Gap Report 2012* (Nairobi: United Nations Environment Programme [UNEP], 2013).

7. D. Archer, "Fate of Fossil Fuel $CO_2$ in Geologic Time," *Journal of Geophysical Research* 110, no. CO9S05 (2005), doi: 10.1029/2004JC002625.

8. Pearson and Palmer, "Atmospheric Carbon Dioxide Concentrations."

9. R. F. Keeling and S. R. Shertz, "Seasonal and Interannual Variations in Atmospheric Oxygen and Implications for the Global Carbon Cycle," *Nature* 358 (1992): 723–727.

10. G. D. Farquhar, J. R. Ehleringer, and K. T. Hubick, "Carbon Isotope Discrimination and Photosynthesis," *Annual Review of Plant Physiology and Plant Molecular Biology* 40 (1989): 503–537.

11. R. F. Keeling, S. C. Piper, A. F. Bollenbacher, and S. J. Walker, *Monthly Atmospheric $^{13}C/^{12}C$ Isotopic Ratios for 11 SIO Stations* (Oak Ridge, TN: Carbon Dioxide Information Analysis Center, Oak Ridge National Laboratory, 2013).

12. S. Kirschke, P. Bousquet, P. Ciais, et al., "Three Decades of Global Methane Sources and Sinks," *Nature Geoscience* 10 (2013): 813–823.

13. A. H. Lockwood, *The Silent Epidemic: Coal and the Hidden Threat to Health* (Cambridge, MA: MIT Press, 2012).

14. A. Karion, C. Sweeney, G. Petron, et al., "Methane Emissions Estimate from Airborne Measurements over a Western United States Natural Gas Field," *Geophysical Research Letters* 40, no. 16 (2013): 4393–4397.

15. N. G. Phillips, R. Ackley, E. R. Crosson, et al., "Mapping Urban Pipeline Leaks: Methane Leaks across Boston," *Environmental Pollution* 173 (2013): 1–4.

16. R. B. Jackson, A. Down, N. G. Phillips, et al., "Natural Gas Pipeline Leaks across Washington, DC," *Environmental Science Technology* 48, no. 3 (2014): 2051–2058.

17. K. L. Mays, P. B. Shepson, B. H. Stirm, et al., "Aircraft-Based Measurements of the Carbon Footprint of Indianapolis," *Environmental Science Technology* 43, no. 20 (2009): 7816–7823; Y-K. Hsu, T. VanCuren, S. Park, et al., "Methane Emissions Inventory Verification in Southern California," *Atmospheric Environment* 44 (2010): 1–7.

18. S. M. Miller, S. C. Wofsy, A. M. Michalak, et al., "Anthropogenic Emissions of Methane in the United States," *Proceedings of the National Academy of Sciences* 110, no. 50 (2013): 20018–20022.

19. A. R. Ingraffea, M. T. Wells, R. L. Santoro, and S. B. C. Shonkoff, "Assessment and Risk Analysis of Casing and Cement Impairment in Oil and Gas Wells in Pennsylvania, 2000–2012," *Proceedings of the National Academy of Sciences* 111, no. 30 (2014): 10955–10960.

20. R. W. Howarth and A. Ingraffea, "Should Fracking Stop?," *Nature* 477 (2011): 271–273.

21. A. Ingraffea, "A Gangplank to a Warm Future," *New York Times*, July 28, 2013.

22. E. A. G. Schurr and B. Abbott, and Permafrost Carbon Network, "High Risk of Permafrost Thaw," *Nature* 480 (2011): 32–33.

23. A. Syakila and C. Kroeze, "The Global Nitrous Oxide Budget Revisited," *Greenhouse Gas Measurement and Management* 1 (2011): 17–26.

24. IPCC, *Fifth Assessment Report of the Intergovernmental Panel on Climate Change* (Geneva: Intergovernmental Panel on Climate Change, 2014).

25. N. Oreskes, "The Scientific Consensus on Climate Change," *Science* 306, no. 5702 (2004): 1686.

## 3 Heat and Severe Weather

1. J. M. Robine, S. L. Cheung, S. Le Roy, et al., "Death Toll Exceeded 70,000 in Europe during the Summer of 2003," *Comptes Rendus Biologies* 331, no. 2 (2008): 171–178.

2. J. Masters, "Over 15,000 Likely Dead in Russian Heat Wave," Weather Underground, August 9, 2010, http://www.wunderground.com/blog/JeffMasters/comment.html?entrynum=1571&tstamp=.

3. T. George, "Pro Football: Heat Kills a Pro Football Player; NFL Orders a Training Review," *New York Times*, August 2, 2001.

4. World Health Organization, "Climate Change and Health: Fact Sheet 266," World Health Organization, 2013.

5. D. Campbell-Lendrum, D. Chadee, Y. Honda, et al., "Human Health: Impacts, Adaptation, and Co-Benefits," in *Climate Change 2014: Impacts, Adaptation, and Vulnerability; Part A: Global and Sectoral Aspects; Contribution of Working Group II to the Fifth Assessment Report of the Intergovernmental Panel on Climate Change*, ed. C. B. Field, V. R. Barros, D. J. Dokken, et al., 709–754 (New York: Cambridge University Press, 2014).

6. See "Annex III: Glossary," in *Climate Change 2013: The Physical Science Basis; Contribution of Working Group I to the Fifth Assessment Report of the Intergovernmental Panel on Climate Change*, ed. T. F. Stocker, D. Qin, G.-K. Plattner, et al., 1447–1466 (New York: Cambridge University Press, 2014).

7. R. Kaiser, A. Le Tetre, J. Schwartz, et al., "The Effect of the 1995 Heat Wave in Chicago on All-Cause and Cause-Specific Mortality," *American Journal of Public Health* 97, Suppl. 1 (2007): S158–S162.

8. National Oceanographic and Atmospheric Administration, *July 1995 Heat Wave: Natural Disaster Report 1995* (Washington, DC: NOAA, 1995).

9. J. C. Semenza, J. E. McCullough, W. D. Flanders, M. A. McGeehin, and J. R. Lumpkin, "Excess Hospital Admissions during the July 1995 Heat Wave in Chicago," *American Journal of Preventive Medicine* 16, no. 4 (1999): 269–277.

10. Kaiser, Le Tetre, Schwartz, et al., "The Effect of the 1995 Heat Wave in Chicago."

11. J. C. Semenza, C. H. Rubin, K. H. Falter, et al., "Heat-Related Deaths during the July 1995 Heat Wave in Chicago," *The New England Journal of Medicine* 335, no. 2 (1996): 84–90.

12. X. Wu, J. E. Brady, H. Rosenberg, and G. Li, "Emergency Department Visits for Heat Stroke in the United States, 2009 and 2010," *Injury Epidemiology* 1, no. 1 (2014): 8–12.

13. Reuters, "India Heatwave: Death Toll Passes 2,500 as Victim Families Fight for Compensation," *The Telegraph*, June 2, 2015.

14. T. Houser, R. Kopp, S. M. Hsiang, et al., *American Climate Prospectus: Economic Risks in the United States* (New York: Rhodium Group, LLC, 2014).

15. A. Barreca, K. Clay, O. Deschênes, et al., "Adapting to Climate Change: The Remarkable Decline in the US Temperature-Mortality Relationship over the 20th Century," NBER Working Paper No. 18692 (Cambridge, MA: National Bureau of Economic Research, 2013).

16. Reuters, "India Heatwave."

17. O. Deschênes and M. Greenstone, "Climate Change, Mortality, and Adaptation: Evidence from Annual Fluctuations in Weather in the US," NBER Working Paper No. 13178 (Cambridge, MA: National Bureau of Economic Research, 2007).

18. R. Basu, W. Y. Feng, and B. D. Ostro, "Characterizing Temperature and Mortality in Nine California Counties," *Epidemiology* 19, no. 1 (2008): 138–145.

19. Y. Honda, M. Kondo, G. McGregor, et al., "Heat-Related Mortality Risk Model for Climate Change Impact Projection," *Environmental Health and Preventative Medicine* 19, no. 1 (2014): 56–63.

20. Campbell-Lendrum, Chadee, Honda, et al., "Human Health: Impacts, Adaptation, and Co-Benefits."

21. P. A. Stott, D. A. Stone, and M. R. Allen, "Human Contribution to the European Heatwave of 2003," *Nature* 432, no. 7017 (2004): 610–614.

22. N. Christidis, P. A. Stott, G. S. Jones, et al., "Human Activity and Anomalously Warm Seasons in Europe," *International Journal of Climatology* 32 (2014): 225–239.

23. Ibid.

24. Semenza, Rubin, Falter, et al., "Heat-Related Deaths during the July 1995 Heat Wave in Chicago."

25. M. Stafoggia, F. Forastiere, P. Michelozzi, and C. A. Perucci, "Summer Temperature-Related Mortality: Effect Modification by Previous Winter Mortality," *Epidemiology* 20, no. 4 (2009): 575–583.

26. C. Carmichael, G. Bickler, S. Kovats, et al., "Overheating and Hospitals—What Do We Know?," *Journal of Hospital Administration* 2, no. 1 (2013): 7.

27. A. Gasparrini and B. Armstrong, "The Impact of Heat Waves on Mortality," *Epidemiology* 22, no. 1 (2011): 68–73.

28. Barreca, Clay, Deschênes, et al., "Adapting to Climate Change."

29. Zahid Arab, "What Caused the Landslide near Oso?," KING 5 News, March 23, 2014, Seattle, WA.

30. P. Y. Groisman, R. W. Knight, and T. R. Karl, "Changes in Intense Precipitation over the Central United States," *Journal of Hydrometeorology* 13 (2012): 47–66.

31. T. R. Karl, J. M. Melillo, and T. C. Peterson, *Global Climate Change Impacts in the United States* (New York: Cambridge University Press, 2009); US Global Change Research Program, *Climate Change Impacts in the United States: The*

*Third National Climate Assessment* (Washington, DC: US Global Change Research Program, 2014).

32. NOAA Hurricane Research Division, "How Much Energy Does a Hurricane Release?," NOAA, 2015, http://www.aoml.noaa.gov/hrd/tcfaq/D7.html.

33. K. E. Kunkel, T. R. Karl, H. Brooks, et al., "Monitoring and Understanding Trends in Extreme Storm: State of Knowledge," *Bulletin of the American Meteorological Society* 94 (2013): 499–514.

34. A. H. Lockwood, *The Silent Epidemic: Coal and the Hidden Threat to Health* (Cambridge, MA: MIT Press, 2012).

35. US Global Change Research Program, *Climate Change Impacts in the United States: The Third National Climate Assessment.*

36. Goddard Earth Sciences and Data Information Services Center, "Hurricane Katrina, August 23–30, 2005," GES DISC, 2005, http://disc.sci.gsfc.nasa.gov/hurricane/additional/science-focus/HurricaneKatrina2005.shtml.

37. S. Fink, *Five Days at Memorial: Life and Death in a Storm-Ravaged Hospital* (New York: Crown Publishers, 2013).

38. H. E. Brooks, "Severe Thunderstorms and Climate Change," *Atmospheric Research* 123 (2013): 129–138.

39. N. S. Diffenbaugh, M. Scherer, and R. J. Trapp, "Robust Increases in Severe Thunderstorm Environments in Response to Greenhouse Forcing," *Proceedings of the National Academy of Sciences* 110, no. 41 (2013): 16361–16366.

## 4   Infectious Diseases

1. S. S. Myers, L. Gaffikin, C. D. Golden, et al., "Human Health Impacts of Ecosystem Alteration," *Proceedings of the National Academy of Sciences* 110, no. 47 (2013): 18753–18760.

2. Ibid.

3. M. C. Thomson, F. J. Doblas-Reyes, S. J. Mason, et al., "Malaria Early Warnings Based on Seasonal Climate Forecasts from Multi-Model Ensembles," *Nature* 439, no. 7076 (2006): 576–579.

4. A. Anyamba, K. J. Linthicum, J. L. Small, et al., "Climate Teleconnections and Recent Patterns of Human and Animal Disease Outbreaks," *PLoS Neglected Tropical Diseases* 6, no. 1 (2012): e1465.

5. World Health Organization, *Handbook for Integrated Vector Management*, 1 (Geneva, Switzerland: WHO, 2012).

6. S. Ranjit and N. Kissoon, "Dengue Hemorrhagic Fever and Shock Syndromes," *Pediatric Critical Care Medicine* 12, no. 1 (2011): 90–100.

7. S. Hales, N. de Wet, J. Maindonald, and A. Woodward, "Potential Effect of Population and Climate Changes on Global Distribution of Dengue Fever: An Empirical Model," *The Lancet* 360, no. 9336 (2002): 830–834.

8. D. Nicks, "Dengue Fever Infections in Florida Make Health Experts Wary of Mosquito-Borne Outbreak," *Time*, June 4, 2014.

9. Ranjit and Kissoon, "Dengue Hemorrhagic Fever and Shock Syndromes."

10. D. Nash, F. Mostashari, A. Fine, et al., "The Outbreak of West Nile Virus Infection in the New York City Area in 1999," *New England Journal of Medicine* 344, no. 24 (2001): 1807–1814.

11. C. W. Morin and A. C. Comrie, "Regional and Seasonal Response of a West Nile Virus Vector to Climate Change," *Proceedings of the National Academy of Sciences* 110, no. 39 (2013): 15620–15625.

12. Ibid.

13. Ibid.

14. F. Cavrini, P. Gaibani, A. M. Pierro, G. Rossini, M. P. Landini, and V. Sambri, "Chikungunya: An Emerging and Spreading Arthropod-Borne Viral Disease," *Journal of Infection in Developing Countries* 3, no. 10 (2009): 744–752.

15. Ibid.

16. L. J. Chang, K. A. Dowd, F. H. Mendoza, et al., "Safety and Tolerability of Chikungunya Virus-Like Particle Vaccine in Healthy Adults: A Phase 1 Dose-Escalation Trial," *The Lancet* 384, no. 9959 (2014): 2046–2052.

17. Centers for Disease Control and Prevention, "Zika Virus," http://www.cdc .gov/zika (January 20, 2016); I. I. Bogoch, O. J. Brady, M. U. G. Kraemer, et al. "Anticipating the International Spread of Zika Virus from Brazil," *The Lancet*, January 14, 2016, http://dx.doi.org/10.1016/ S0140-6736(16)00080-5.

18. Centers for Disease Control and Prevention, "Lyme Disease," CDC, 2015, http://www.cdc.gov/lyme.

19. M. G. Morshed, J. D. Scott, K. Fernando, et al., "Migratory Songbirds Disperse Ticks across Canada, and First Isolation of the Lyme Disease Spirochete, *Borrelia burgdorferi*, from the Avian Tick, *Ixodes auritulus*," *The Journal of Parasitology* 91, no. 4 (2005): 780–790.

20. N. H. Ogden, A. Maarouf, I. K. Barker, et al., "Climate Change and the Potential for Range Expansion of the Lyme Disease Vector *Ixodes scapularis* in Canada," *International Journal for Parasitology* 36, no. 1 (2006): 63–70.

21. R. Lozano, M. Naghavi, K. Foreman, et al., "Global and Regional Mortality from 235 Causes of Death for 20 Age Groups in 1990 and 2010: A Systematic Analysis for the Global Burden of Disease Study 2010," *The Lancet* 380, no. 9859 (2012): 2095–2128.

22. World Health Organization, *World Malaria Report 2013* (Geneva: WHO, 2013).

23. C. J. Murray, L. C. Rosenfeld, S. S. Lim, et al., "Global Malaria Mortality between 1980 and 2010: A Systematic Analysis," *The Lancet* 379, no. 9814 (2012): 413–431.

24. P. McKenna, "Nobel Prize Goes to Modest Woman Who Beat Malaria for China," *New Scientist*, October 5, 2015.

25. K. A. Cullen and P. M. Arguin, *Morbidity and Mortality Weekly Report, Malaria Surveillance—United States, 2012* (Atlanta: Centers for Disease Control and Prevention, 2014).

26. Centers for Disease Control and Prevention, "Malaria Parasites," CDC, 2015, http://www.cdc.gov/malaria.

27. Myers, Gaffikin, Golden, et al., "Human Health Impacts of Ecosystem Alteration."

28. D. Campbell-Lendrum, D. Chadee, Y. Honda, et al., "Human Health: Impacts, Adaptation, and Co-Benefits," in *Climate Change 2014: Impacts, Adaptation, and Vulnerability; Part A: Global and Sectoral Aspects; Contribution of Working Group II to the Fifth Assessment Report of the Intergovernmental Panel on Climate Change*, ed. C. B. Field, V. R. Barros, D. J. Dokken, et al., 709–754 (New York: Cambridge University Press, 2014).

29. World Health Organization, *World Malaria Report 2013*.

30. Campbell-Lendrum, Chadee, Honda, et al., "Human Health: Impacts, Adaptation, and Co-Benefits."

31. M. Karanikolos, P. Mladovsky, J. Cylus, et al., "Financial Crisis, Austerity, and Health in Europe," *The Lancet* 381, no. 9874 (2013): 1323–1331.

32. Campbell-Lendrum, Chadee, Honda, et al., "Human Health: Impacts, Adaptation, and Co-Benefits."

33. S. Paz, "Impact of Temperature Variability on Cholera Incidence in Southeastern Africa, 1971–2006," *Ecohealth* 6, no. 3 (2009): 340–345.

34. M. S. Islam, M. A. Sharker, S. Rheman, et al., "Effects of Local Climate Variability on Transmission Dynamics of Cholera in Matlab, Bangladesh," *Transactions of the Royal Society of Tropical Medicine and Hygiene* 103, no. 11 (2009): 1165–1170.

35. R. Reyburn, D. R. Kim, M. Emch, et al., "Climate Variability and the Outbreaks of Cholera in Zanzibar, East Africa: A Time Series Analysis," *The American Journal of Tropical Medicine and Hygiene* 84, no. 6 (2011): 862–869.

36. World Health Organization, "Weekly Epidemiological Record: Cholera 2012," *Weekly Epidemiological Record* 88, no. 31 (2013): 321–336.

37. A. L. C. F. Cravioto, D. S. Lantagne, and G. B. Nair, *Final Report of the Independent Panel of Experts on the Cholera Outbreak in Haiti*, United Nations, 2011, http://www.un.org/News/dh/infocus/haiti/UN-cholera-report-final.pdf.

38. Pan American Health Organization, World Health Organization, "Cholera," *Epidemiological Update*, June 27, 2014.

39. H. W. Lee, "Hemorrhagic Fever with Renal Syndrome in Korea," *Reviews of Infectious Diseases* 11, Supplement 4 (1989): S864–S876.

40. J. D. Boone, K. C. McGwire, E. W. Otteson, et al., "Remote Sensing and Geographic Information Systems: Charting Sin Nombre Virus Infections in Deer Mice," *Emerging Infectious Diseases* 6, no. 3 (2000): 248–258; G. E. Glass, J. E.

Cheek, J. A. Patz, et al., "Using Remotely Sensed Data to Identify Areas at Risk for Hantavirus Pulmonary Syndrome," *Emerging Infectious Diseases* 6, no. 3 (2000): 238–247.

41. C. Gonzalez, O. Wang, S. E. Strutz, C. Gonzalez-Salazar, V. Sanchez-Cordero, and S. Sarkar, "Climate Change and Risk of Leishmaniasis in North America: Predictions from Ecological Niche Models of Vector and Reservoir Species," *PLoS Neglected Tropical Diseases* 4, no. 1 (2010): e585.

42. M. R. V. Sant'Anna, H. Diaz-Albiter, K. Aguiar-Martins, et al., "Colonisation Resistance in the Sand Fly Gut: Leishmania Protects *Lutzomyia longipalpis* from Bacterial Infection," *Parasites and Vectors* 7 (2014): 329.

## 5  Climate Change, Agriculture, and Famine

1. Bread for the World Institute, "Hunger Report: Ending Hunger in America," Bread for the World, September 14, 2014.

2. International Food Policy Research Institute, *Global Nutrition Report 2015: Actions and Accountability to Advance Nutrition and Sustainable Development* (Washington, DC: IFPRI, 2015).

3. R. K. Berner and Z. Kothavala, "GEOCARB III: A Revised Model of Atmospheric $CO_2$ over Phanerozoic Time," *American Journal of Science* 301 (2001): 182–204.

4. P. N. Pearson and M. R. Palmer, "Atmospheric Carbon Dioxide Concentrations over the Past 60 Million Years," *Nature* 406, no. 6797 (2000): 695–699.

5. C. L. Walthall, P. Hatfield, L. Backlund, et al., *Climate Change and Agriculture in the United States: Effects and Adaptation; USDA Technical Bulletin 1935* (Washington, DC: United States Department of Agriculture, 2012).

6. H. W. Polley, H. B. Johnson, and J. D. Derner, "Increasing $CO_2$ from Subambient to Superambient Concentrations Alters Species Composition and Increases Above-Ground Biomass in a $C_3/C_4$ Grassland," *New Phytologist* 160, no. 2 (2003): 319–327.

7. L. H. Ziska and J. A. Bunce, "Predicting the Impact of Changing $CO_2$ on Crop Yields: Some Thoughts on Food," *New Phytologist* 175, no. 4 (2007): 607–618.

8. R. J. W. Brienen, O. L. Phillips, T. R. Feldpausch, et al., "Long-Term Decline of the Amazon Carbon Sink," *Nature* 519, no. 7543 (2015): 344–348.

9. Ibid., 344.

10. E. Kolbert, *The Sixth Extinction: An Unnatural History* (New York: Henry Holt & Co., 2014).

11. P. B. Reich, S. E. Hobbie, T. Lee, et al., "Nitrogen Limitation Constrains Sustainability of Ecosystem Response to $CO_2$," *Nature* 440, no. 7086 (2006): 922–925.

12. Ibid.; S. S. Myers, A. Zanobetti, I. Kloog, et al., "Increasing $CO_2$ Threatens Human Nutrition," *Nature* 510, no. 7503 (2014): 139–142.

13. C. A. Rogers, P. M. Wayne, E. A. Macklin, et al., "Interaction of the Onset of Spring and Elevated Atmospheric $CO_2$ on Ragweed (*Ambrosia artemisiifolia* L.) Pollen Production," *Environmental Health Perspectives* 114 (2006): 865–869.

14. Walthall Hatfield, Backlund, et al., *Climate Change and Agriculture in the United States.*

15. W. Schlenker and M. J. Roberts, "Nonlinear Temperature Effects Indicate Severe Damages to U.S. Crop Yields under Climate Change," *Proceedings of the National Academy of Sciences* 106, no. 37 (2009): 15594–15598.

16. Schlenker and Roberts, "Nonlinear Temperature Effects."

17. D. B. Lobell, W. Schlenker, and J. Costa-Roberts, "Climate Trends and Global Crop Production since 1980," *Science* 333 (2011): 616–620.

18. Walthall Hatfield, Backlund, et al., *Climate Change and Agriculture in the United States.*

19. J. Diamond, *Collapse: How Societies Choose to Fail or Succeed* (New York: Viking, 2005).

20. L. A. Javier, "U.S. Drought Persisting Seen as Threat to Corn, Soybeans," *Bloomberg*, January 9, 2013.

21. N. Cumming-Bruce, "South Sudan Urgently Needs Help to Stave Off Famine, U.N. Warns," *New York Times*, April 3, 2014.

22. M. Ravallion, *Famines and Economics*, n.p. (Washington, DC: World Bank Group, 1996).

23. IPCC, "Summary for Policymakers," in *Climate Change 2014: Impacts, Adaptation, and Vulnerability; Part A: Global and Sectoral Aspects; Contribution of Working Group II to the Fifth Assessment Report of the Intergovernmental Panel on Climate Change*, ed. C. B. Field, V. R. Barros, D. J. Dokken, et al. (New York: Cambridge University Press, 2014), 1–32.

24. FAO, IFAD, and WFP, *The State of Food Insecurity in the World 2014: Strengthening the Enabling Environment for Food Security and Nutrition* (Rome: FAO, 2014).

25. M. Lagi, K. Z. Bertrand, and Y. Bar-Yam, "The Food Crises and Political Instability in North Africa and the Middle East," 2011, http://papers.ssrn.com/sol3/papers.cfm?abstract_id=1910031.

26. T. R. Malthus, *An Essay on the Principle of Population, as It Affects the Future Improvement of Society. With Remarks on the Speculations of Mr. Godwin, M. Condorcet and Other Writers* (London: J. Johnson, London, 1798).

# 6   Sea Level Rise and Environmental Refugees

1. P. Chapman, "Entire Nation of Kiribati to be Relocated over Rising Sea Level Threat," *The Telegraph*, March 7, 2012.

2. G. McGranahan, D. Balk, and B. Anderson, "The Rising Tide: Assessing the Risks of Climate Change and Human Settlements in Low Elevation Coastal Zones," *Environment and Urbanization* 19, no. 1 (2007): 17–37.

3. See "Annex III: Glossary," in *Climate Change 2013: The Physical Science Basis; Contribution of Working Group I to the Fifth Assessment Report of the Intergovernmental Panel on Climate Change*, ed. T. F. Stocker, D. Qin, G.-K. Plattner, et al., 1447–1466 (New York: Cambridge University Press, 2014).

4. J. Martinich, J. Neumann, L. Ludwig, and L. Jantarasami, "Risks of Sea Level Rise to Disadvantaged Communities in the United States," *Mitigation and Adaptation Strategies for Global Change* 18, no. 2 (2013): 169–185.

5. C. S. Fulthorpe, K. G. Miller, A. W. Droxler, S. P. Hesselbo, G. F. Camoin, and M. A. Kominz, "Drilling to Decipher Long-Term Sea-Level Changes and Effects— A Joint Consortium for Ocean Leadership, ICDP, IODDP, DOSECC, and Chevron Workshop," *Scientific Drilling* 6 (2008): 19–28.

6. J. A. Church and N. J. White, "A 20th Century Acceleration in Global Sea-Level Rise," *Geophysical Research Letters* 33, no. 1 (2006): L01602.

7. A. Cazenave and R. S. Nerem, "Present-Day Sea Level Change: Observations and Causes," *Reviews of Geophysics* 42, no. 3 (2004): RG3001.

8. J. A. Church, P. U. Clark, A. Cazenave, et al., "Sea Level Change," in *Climate Change 2013: The Physical Science Basis; Contribution of Working Group I to the Fifth Assessment Report of the Intergovernmental Panel on Climate Change*, ed. T. F. Stocker, D. Qin, G-K. Plattner, et al., 1137–1216 (New York: Cambridge University Press, 2014).

9. Ibid.

10. Ibid.; N. Bouttes, J. M. Gregory, and J. A. Lowe, "The Reversibility of Sea Level Rise," *Journal of Climate* 26 (2013): 2502–2513.

11. Bouttes, Gregory, and Lowe, "The Reversibility of Sea Level Rise," 2502.

12. Church, Clark, Cazenave, et al., "Sea Level Change."

13. E. Rignot, J. Mouginot, M. Morlighem, H. Seroussi, and B. Scheuchl, "Widespread, Rapid Grounding Line Retreat of Pine Island, Thwaites, Smith, and Kohler Glaciers, West Antarctica, from 1992 to 2011," *Geophysical Research Letters* 41, no. 10 (2014): 3502–3509; I. Joughin, B. E. Smith, and B. Medley, "Marine Ice Sheet Collapse Potentially Under Way for the Thwaites Glacier Basin, West Antarctica," *Science* 344, no. 6185 (2014): 735–738.

14. R. Winkelmann, A. Levermann, A. Ridgwell, and K. Caldeira, "Combustion of Available Fossil Fuel Resources Sufficient to Eliminate the Antarctic Ice Sheet," *Science Advances* 1, no. 8 (2015): e1500589.

15. B. F. Chao, Y. H. Wu, and Y. S. Li, "Impact of Artificial Reservoir Water Impoundment on Global Sea Level," *Science* 320, no. 5873 (2008): 212–214.

16. B. Henson, "Dissecting Sandy's Surge," NCAR UCAR *AtmosNews*, December 31, 2012.

17. Ibid.

18. "Hurricane Sandy," *Wikipedia, The Free Encyclopedia*, accessed December 19, 2015, https://en.wikipedia.org/w/index.php?title=Hurricane_Sandy&oldid=696766390.

19. W. Sullivan, "Cyclone May Be the Worst Catastrophe of Century," *New York Times*, November 22, 1970.

20. D. Campbell-Lendrum, D. Chadee D, Y. Honda Y, et al., "Human Health: Impacts, Adaptation, and Co-Benefits," in *Climate Change 2014: Impacts, Adaptation, and Vulnerability; Part A: Global and Sectoral Aspects; Contribution of Working Group II to the Fifth Assessment Report of the Intergovernmental Panel on Climate Change*, ed. C. B. Field, V. R. Barros, D. J. Dokken, et al., 709–754 (New York: Cambridge University Press, 2014).

21. U. Haque, M. Hashizume, K. N. Kolivras, H. J. Overgaard, B. Das, and T. Yamamoto, "Reduced Death Rates from Cyclones in Bangladesh: What More Needs to Be Done?," *Bulletin of the World Health Organization* 90, no. 2 (2012): 150–156.

22. B. Copeland, J. Keller, and B. Marsh, "What Could Disappear," *New York Times*, November 24, 2012.

23. E. Genovese, S. Hallegatte, and P. Dumas, "Damage Assessment from Storm Surge to Coastal Cities: Lessons from the Miami Area," in *Advancing Geoinformation Science for a Changing World*, ed. S. Geertman, W. Reinhardt, and F. Toppen, 21–43 (Toronto: University of Toronto Press, 2011).

24. S. Hanson, R. Nicholls, N. Ranger, et al., "A Global Ranking of Port Cities with High Exposure to Climate Extremes," *Climatic Change* 104, no. 1 (2011): 89–111.

25. S. Hallegatte, N. Ranger, O. Mestre, et al., "Assessing Climate Change Impacts, Sea Level Rise and Storm Surge Risk in Port Cities: A Case Study on Copenhagen," *Climatic Change* 104, no. 1 (2011): 113–137.

26. R. A. Pilke Jr., J. Gratz, C. W. Landsea, D. Collins, M. A. Saunders, and R. Musulin, "Normalized Hurricane Damage in the United States: 1900–2005," *Natural Hazards Review* 9, no. 1 (2008): 29–42.

27. M. R. Besonen, R. S. Bradley, M. Mudelsee, and M. B. Abbott, "A 1,000 Year, Annually-Resolved Record of Hurricane Activity from Boston, Massachusetts," *Geophysical Research Letters* 35, no. 14 (2014): L14705.

28. M. A. Bender, T. R. Knutson, R. E. Tuleya, et al., "Modeled Impact of Anthropogenic Warming on the Frequency of Intense Atlantic Hurricanes," *Science* 327, no. 5964 (2010): 454–458.

29. T. R. Knutson, J. L. McBride, J. Chan, et al., "Tropical Cyclones and Climate Change," *Nature Geoscience* 3, no. 3 (2010): 157–163.

## 7   Air Pollution, Air Quality, and Climate Change

1. J. Evelyn, *Fumifugium: The Inconvenience of the Aer and Smoak of London; Together with some Remedies Humbly Proposed by J. E. Esq; To His Sacred Majestie and to the Parliament now Assembled* (London: Published by His Majesties Command, 2011).

2. B. Z. Simkhovich, M. T. Kleinman, and R. A. Kloner, "Air Pollution and Cardiovascular Injury Epidemiology, Toxicology, and Mechanisms," *Journal of the American College of Cardiology* 52, no. 9 (2008): 719–726.

3. R. L. Canfield, C. R. Henderson Jr., D. A. Cory-Slechta, C. Cox, T. A. Jusko, and B. P. Lanphear, "Intellectual Impairment in Children with Blood Lead Concentrations below 10 µg per Deciliter," *New England Journal of Medicine* 348, no. 16 (2003): 1517–1526.

4. US Environmental Protection Agency, Office of Air and Radiation, *The Benefits and Costs of the Clean Air Act from 1990 to 2020* (Washington, DC: EPA, 2011).

5. J. Allen, "Chemistry in the Sunlight," *NASA Earth Observatory*, January 27, 2002, http://earthobservatory.nasa.gov/Features/ChemistrySunlight.

6. A. Arneth, R. K. Monson, G. Schurgers, Ü. Niinemets, and P. I. Palmer, "Why Are Estimates of Global Terrestrial Isoprene Emissions So Similar (and Why Is This Not So for Monoterpenes)?," *Atmospheric Chemistry and Physics* 8 (2008): 4605–4620.

7. T. F. Stocker, D. Qin, G-K. Plattner, et al., eds., *Climate Change 2013: The Physical Science Basis; Contribution of Working Group I to the Fifth Assessment Report of the Intergovernmental Panel on Climate Change* (New York: Cambridge University Press, 2014).

8. Centers for Disease Control and Prevention, "Asthma Surveillance Data, 2013," 2015, http://www.cdc.gov/asthma/asthmadata.htm.

9. US Environmental Protection Agency, "Regulatory Actions: Ground-Level Ozone," EPA, December 10, 2014, http://www3.epa.gov/ttn/naaqs/standards/ozone/s_o3_index.html.

10. IPCC, *Fifth Assessment Report of the Intergovernmental Panel on Climate Change* (Geneva: Intergovernmental Panel on Climate Change, 2014).

11. US Environmental Protection Agency, "Regulatory Actions: Ground-Level Ozone."

12. Ibid.

13. W. C. Adams, "Comparison of Chamber and Face-Mask 6.6-Hour Exposures to Ozone on Pulmonary Function and Symptoms Responses," *Inhalation Toxicology* 14, no. 7 (2002): 745–764; W. C. Adams, "Comparison of Chamber 6.6-h Exposures to 0.04–0.08 PPM Ozone via Square-Wave and Triangular Profiles on Pulmonary Responses," *Inhalation Toxicology* 18, no. 2 (2006): 127–136; J. E. Goodman, R. L. Prueitt, J. Chandalia, and S. N. Sax, "Evaluation of Adverse Human Lung Function Effects in Controlled Ozone Exposure Studies," *Journal of Applied Toxicology* 34, no. 5 (2014): 516–524; E. S. Schelegle, C. A. Morales, W. F. Walby, et al., "6.6-Hour Inhalation of Ozone Concentrations from 60 to 87 Parts per Billion in Healthy Humans," *American Journal of Respiratory and Critical Care Medicine* 180, no. 3 (2009): 265–272.

14. J. S. Brown, T. F. Bateson, and W. F. McDonnell, "Effects of Exposure to 0.06 ppm Ozone on FEV1 in Humans: A Secondary Analysis of Existing Data," *Environmental Health Perspectives* 116, no. 8 (2008): 1023–1026.

15. C. S. Kim, N. E. Alexis, A. G. Rappold, et al., "Lung Function and Inflammatory Responses in Healthy Young Adults Exposed to 0.06 ppm Ozone for 6.6 Hours," *American Journal of Respiratory Critical Care Medicine* 183, no. 9 (2011): 1215–1221.

16. C. H. Chen, C. C. Chan, B. Y. Chen, T. J. Cheng, and Y. L. Guo, "Effects of Particulate Air Pollution and Ozone on Lung Function in Non-asthmatic Children," *Environmental Research* 137 (2015): 40–48.

17. S. Wu, L. J. Mickley, D. J. Jacob, D. Rind, and D. G. Streets, "Effects of 2000–2050 Changes in Climate and Emissions on Global Tropospheric Ozone and the Policy-Relevant Background Surface Ozone in the United States," *Journal of Geophysical Research: Atmospheres* 113, no. D18 (2008): D18312.

18. M. L. Bell, R. Goldberg, C. Hogrefe, et al., "Climate Change, Ambient Ozone and Health in 50 US Cities," *Climatic Change* 82, no. 61 (2007): 76.

19. E. Tagaris, K. Manomaiphiboon, K. J. Liao, et al., "Impacts of Global Climate Change and Emissions on Regional Ozone and Fine Particulate Matter Concentrations over the United States," *Journal of Geophysical Research: Atmospheres* 112, no. D14 (2007): D14312.

20. C. Warneke, J. A. de Gouw, A. Stohl, et al., "Biomass Burning and Anthropogenic Sources of CO over New England in the Summer 2004," *Journal of Geophysical Research: Atmospheres* 111, no. D23 (2006): D23S15.

21. S. Wu, L. J. Mickley, E. M. Leibensperger, D. J. Jacob, D. Rind, and D. G. Streets, "Effects of 2000–2050 Global Change on Ozone Air Quality in the United States," *Journal of Geophysical Research: Atmospheres* 113, no. D6 (2008): D06302.

22. Ibid.

23. A. H. Lockwood, *The Silent Epidemic: Coal and the Hidden Threat to Health* (Cambridge, MA: MIT Press, 2012).

24. D. J. Jacob and D. A. Winner, "Effect of Climate Change on Air Quality," *Atmospheric Environment* 43, no. 1 (2009): 51–63.

25. Ibid.

26. E. K. Wise and A. C. Comrie, "Meteorologically Adjusted Urban Air Quality Trends in the Southwestern United States," *Atmospheric Environment* 39, no. 16 (2005): 2969–2980.

27. Ibid.

28. Y. J. Balkanski, D. J. Jacob, G. M. Gardner, et al., "Transport and Residence Times of Tropospheric Aerosols Inferred from a Global Three-Dimensional Simulation of $^{210}$Pb," *Journal of Geophysical Research: Atmospheres* 98, no. D11 (1993): 20573–20586.

29. Jacob and Winner, "Effect of Climate Change on Air Quality."

30. J. M. Melillo, T. C. Richmond, and G. W. Yohe, eds., *Climate Change Impacts in the United States: The Third National Climate Assessment* (Washington, DC: US Global Change Research Program, 2014).

31. R. Vautard, M. Beekmann, J. Desplat, A. Hodzic, and S. Morel, "Air Quality in Europe during the Summer of 2003 as a Prototype of Air Quality in a Warmer Climate," *Comptes Rendus Geoscience* 339, no. 11 (2007): 747–763.

32. R. J. Delfino, S. Brummel, J. Wu, et al., "The Relationship of Respiratory and Cardiovascular Hospital Admissions to the Southern California Wildfires of 2003," *Occupational and Environmental Medicine* 66, no. 3 (2009): 189–197.

33. G. Oberdorster, Z. Sharp, V. Atudorei, et al., "Translocation of Inhaled Ultrafine Particles to the Brain," *Inhalation Toxicology* 16, nos. 6–7 (2004): 437–445.

34. L. Calderon-Garciduenas, B. Azzarelli, H. Acuna, et al., "Air Pollution and Brain Damage," *Toxicologic Pathology* 30, no. 3 (2002): 373–389; L. Calderon-Garciduenas, W. Reed, R. R. Maronpot, et al., "Brain Inflammation and Alzheimer's-Like Pathology in Individuals Exposed to Severe Air Pollution," *Toxicologic Pathology* 32, no. 6 (2004): 650–658.

35. A. Nel, T. Xia, L. Madler, and N. Li, "Toxic Potential of Materials at the Nanolevel," *Science* 311, no. 5761 (2006): 622–627.

36. Lockwood, *The Silent Epidemic*.

37. F. Laden, L. M. Neas, D. W. Dockery, and J. Schwartz, "Association of Fine Particulate Matter from Different Sources with Daily Mortality in Six U.S. Cities," *Environmental Health Perspectives* 108, no. 10 (2000): 941–947.

38. I. P. O'Connor and J. C. Wenger, "Particulate Pollution in Ireland from Solid Fuel Burning," paper presented at the Asthma Society Meeting, Killarney, December 3, 2014.

39. G. A. Wellenius, M. R. Burger, B. A. Coull, et al., "Ambient Air Pollution and the Risk of Acute Ischemic Stroke," *Archives of Internal Medicine* 172, no. 3 (2012): 229–234.

40. Lockwood, *The Silent Epidemic*.

# 8  Violence, Conflict, and Societal Disruption

1. S. M. Hsiang, M. Burke, and E. Miguel, "Quantifying the Influence of Climate on Human Conflict," *Science* 341, no. 6151 (2013): 1235367. DOI: 10.1126/science.1235367.

2. A. Reifman, R. Larrick, and S. Fein, "Temper and Temperature on the Diamond: The Heat-Aggression Relationship in Major League Baseball," *Personality and Social Psychology Bulletin* 17, no. 5 (1991): 580–585.

3. R. P. Larrick, T. A. Timmerman, A. M. Carton, and J. Abrevaya, "Temper, Temperature, and Temptation Heat-Related Retaliation in Baseball," *Psychological Science* 22, no. 4 (2011): 423–428.

4. D. Card and G. B. Dahl, "Family Violence and Football: The Effect of Unexpected Emotional Cues on Violent Behavior," *The Quarterly Journal of Economics* 126, no. 1 (2011): 103–143.

5. D. T. Kenrick and S. W. MacFarlane, "Ambient Temperature and Horn Honking: A Field Study of the Heat/Aggression Relationship, *Environment and Behavior* 18, no. 2 (1986): 179–191.

6. A. Vrij, J. Van der Steen, and L. Koppelaar, "Aggression of Police Officers as a Function of Temperature: An Experiment with the Fire Arms Training System," *Journal of Community and Applied Social Psychology* 4, no. 5 (1994): 365–370.

7. C. A. Anderson, W. E. Deuser, and K. M. DeNeve, "Hot Temperatures, Hostile Affect, Hostile Cognition, and Arousal: Tests of a General Model of Affective Aggression," *Personality and Social Psychology Bulletin* 21, no. 5 (1995): 434–448.

8. C. A. Anderson, K. B. Anderson, N. Dorr, et al., "Temperature and Aggression," *Advances in Experimental Social Psychology* 32 (2000): 63–133.

9. C. A. Anderson, "Heat and Violence," *Current Directions in Psychological Science* 10, no. 1 (2001): 33–38.

10. Anderson, Deuser, and DeNeve, "Hot Temperatures, Hostile Affect, Hostile Cognition, and Arousal."

11. D. Kahneman, *Thinking, Fast and Slow* (New York: Farrar, Straus and Giroux, New York, 2011).

12. Anderson, Anderson, Dorr, et al., "Temperature and Aggression."

13. C. S. Hendrix and I. Salehyan, "Climate Change, Rainfall, and Social Conflict in Africa," *Journal of Peace Research* 49, no. 1 (2012): 35–50.

14. Ibid.

15. S. M. Hsiang, K. C. Meng, and M. A. Cane, "Civil Conflicts Are Associated with the Global Climate," *Nature* 476, no. 7361 (2011): 438–441.

16. Ibid.

17. N. Watts, W. N. Adger, P. Agnolucci, et al., "Health and Climate Change: Policy Responses to Protect Public Health," *The Lancet*, June 23, 2015, http:// dx.doi.org/10.1016/S0140-6736(15)60854-6.

18. M. Lagi, K. Z. Bertrand, and Y. Bar-Yam, "The Food Crises and Political Instability in North Africa and the Middle East," 2011, http://papers.ssrn.com/sol3/papers.cfm?abstract_id=1910031.

19. J. Barbet-Gross and J. Cuesta, "Food Riots: From Definition to Operationalization," World Bank Group, undated online note, http://www.worldbank.org/content/dam/Worldbank/document/Poverty%20documents/Introduction%20Guide%20for%20the%20Food%20Riot%20Radar.pdf.

20. J. Diamond, *Collapse: How Societies Choose to Fail or Succeed* (New York: Viking, 2005).

21. Hsiang, Burke, and Miguel, "Quantifying the Influence of Climate on Human Conflict."

22. A. E. Douglass, "The Secret of the Southwest Solved by Talkative Tree Rings," *National Geographic*, December 1929, 736–770.

23. Ibid., 741 (emphasis in original).

24. Ibid., 743.

25. Diamond, *Collapse*.

26. R. L. Kelly, T. A. Surovell, B. N. Shuman, and G. M. Smith, "A Continuous Climatic Impact on Holocene Human Population in the Rocky Mountains," *Proceedings of the National Academy of Sciences* 110, no. 2 (2013): 443–447.

27. B. M. Buckley, K. J. Anchukaitis, D. Penny, et al., "Climate as a Contributing Factor in the Demise of Angkor, Cambodia," *Proceedings of the National Academy of Sciences* 107, no. 15 (2010): 6748–6752.

28. IPCC, *Fifth Assessment Report of the Intergovernmental Panel on Climate Change* (Geneva: Intergovernmental Panel on Climate Change, 2014).

29. D. P. van Vuuren, J. Edmonds, M. Kainuma, et al., "The Representative Concentration Pathways: An Overview," *Climatic Change* 109, nos. 1–2 (2011): 5–31.

30. Hsiang, Burke, and Miguel, "Quantifying the Influence of Climate on Human Conflict."

31. Hsiang, Meng, and Cane, "Civil Conflicts Are Associated with the Global Climate."

## 9   Economic Considerations of Climate Change and Health

1. M. Ruth, D. Coelho, and D. Karetnikov, *The US Economic Impacts of Climate Change and the Costs of Inaction* (College Park: Center for Integrative Environmental Research, University of Maryland, 2007).

2. Ibid.

3. S. Hallegatte, M. Chambwera, G. Heal, et al., "Economics of Adaptation," in *Climate Change 2014: Impacts, Adaptation, and Vulnerability; Part A: Global and Sectoral Aspects; Contribution of Working Group II to the Fifth Assessment Report of the Intergovernmental Panel on Climate Change*, ed. C. B. Field, V. R. Barros, D. J. Dokken, et al., 945–978 (New York: Cambridge University Press, 2014).

4. M. L. Parry, N. Arnell, P. Berry, et al., *Assessing the Costs of Adaptation to Climate Change: A Review of the UNFCCC and Other Recent Estimates*, 8 (London: International Institute for Environment and Development, 2009).

5. T. Houser, R. Kopp, S. M. Hsiang, et al., *American Climate Prospectus: Economic Risks in the United States* (New York: Rhodium Group, LLC, 2014).

6. D. Bonauto, E. Rauser, and L. Lim, "Occupational Heat Illness in Washington State, 2000–2009," Washington State Department of Labor & Industries, 2010, http://www.lni.wa.gov/safety/research/files/occheatrelatedillnesswa20002009.pdf.

7. C. T. Merrill, M. Miller, and C. Steiner, "Hospital Stays Resulting from Excessive Heat and Cold Exposure Due to Weather Conditions in US Community Hospitals, 2005," July 2008, http://www.ncbi.nlm.nih.gov/books/NBK56045.

8. K. Knowlton, M. Rotkin-Ellman, G. King, et al., "The 2006 California Heat Wave: Impacts on Hospitalizations and Emergency Department Visits," *Environmental Health Perspectives* 117, no. 1 (2009): 61–67.

9. R. Lozano, M. Naghavi, K. Foreman, et al., "Global and Regional Mortality from 235 Causes of Death for 20 Age Groups in 1990 and 2010: A Systematic Analysis for the Global Burden of Disease Study 2010," *The Lancet* 380, no. 9859 (2012): 2095–2128.

10. US Environmental Protection Agency, Office of Air and Radiation, *The Benefits and Costs of the Clean Air Act from 1990 to 2020* (Washington, DC: EPA, 2011).

11. D. J. Gubler, "Epidemic Dengue/Dengue Hemorrhagic Fever as a Public Health, Social and Economic Problem in the 21st Century," *Trends in Microbiology* 10, no. 2 (2002): 100–103.

12. D. J. Gubler, "The Economic Burden of Dengue," *The American Journal of Tropical Medicine and Hygiene* 86, no. 5 (2012): 743–744.

13. Y. A. Halasa, D. S. Shepard, and W. Zeng, "Economic Cost of Dengue in Puerto Rico," *The American Journal of Tropical Medicine and Hygiene* 86, no. 5 (2012): 745–752.

14. D. S. Shepard, Y. A. Halasa, B. K. Tyagi, et al., "Economic and Disease Burden of Dengue Illness in India," *The American Journal of Tropical Medicine and Hygiene* 91, no. 6 (2014): 1235–1242.

15. Gubler, "Epidemic Dengue/Dengue Hemorrhagic Fever as a Public Health, Social and Economic Problem."

16. Gubler, "The Economic Burden of Dengue."

17. M. E. Beatty, P. Beutels, M. I. Meltzer, et al., "Health Economics of Dengue: A Systematic Literature Review and Expert Panel's Assessment," *The American Journal of Tropical Medicine and Hygiene* 84, no. 3 (2011): 473–488.

18. Beatty et al, "Health Economics of Dengue."

19. L. Villar, G. H. Dayan, J. L. Arredondo-García, et al., "Efficacy of a Tetravalent Dengue Vaccine in Children in Latin America," *New England Journal of Medicine* 372, no. 2 (2015): 113–123.

20. J. Sachs and P. Malaney, "The Economic and Social Burden of Malaria," *Nature* 415, no. 6872 (2002): 680–685.

21. S. Russell, "The Economic Burden of Illness for Households in Developing Countries: A Review of Studies Focusing on Malaria, Tuberculosis, and Human Immunodeficiency Virus/Acquired Immunodeficiency Syndrome," *The American Journal of Tropical Medicine and Hygiene* 71, suppl. 2 (2004): 147–155.

22. Sachs and Malaney, "The Economic and Social Burden of Malaria."

23. Ibid.

24. Ibid.

25. K. Gordon, G. Lewis, and J. Rogers, "Risky Business: The Economic Risks of Climate Change in the United States," *riskybusiness.org*, Risky Business Project,

June 2014, http://riskybusiness.org/site/assets/uploads/2015/09/RiskyBusiness _Report_WEB_09_08_14.pdf.

26. G. Yohe and M. Schlesinger, "Sea-Level Change: The Expected Economic Cost of Protection or Abandonment in the United States," *Climatic Change* 38, no. 4 (1998): 447–472.

27. J. Neumann, K. Emanuel, S. Ravela, et al., "Joint Effects of Storm Surge and Sea-Level Rise on US Coasts: New Economic Estimates Of Impacts, Adaptation, and Benefits of Mitigation Policy," *Climatic Change* 129, nos. 1–2 (2015): 337–349.

28. J. Martinich, J. Neumann, L. Ludwig, and L. Jantarasami, "Risks of Sea Level Rise to Disadvantaged Communities in the United States," *Mitigation and Adaptation Strategies for Global Change* 18, no. 2 (2013): 169–185.

29. J. L. Hatfield, R. M. Cruse, and M. D. Tomer, "Convergence of Agricultural Intensification and Climate Change in the Midwestern United States: Implications for Soil and Water Conservation," *Marine and Freshwater Research* 64, no. 5 (2013): 423–435.

30. Houser, Kopp, Hsiang, et al., *American Climate Prospectus.*

31. Ibid.

32. US Office of Justice Programs, Bureau of Justice Statistics, 2015, http://www .bjs.gov.

33. Houser, Kopp, Hsiang, et al., *American Climate Prospectus.*

34. M. Ranson, "Crime, Weather, and Climate Change," *Journal of Environmental Economics and Management* 67, no. 3 (2014): 274–302.

35. B. Jacob, L. Lefgren, and E. Moretti, "The Dynamics of Criminal Behavior Evidence from Weather Shocks," *Journal of Human Resources* 42, no. 3 (2007): 489–527.

## 10  Protecting Health

1. J. Steinbeck, *The Grapes of Wrath* (New York: Viking Press, 1939).

2. M. Parry, J. Lowe, and C. Hanson, "Overshoot, Adapt and Recover," *Nature* 458, no. 7242 (2009): 1102–1103.

3. S. S. Lim, T. Vos, A. D. Flaxman, et al., "A Comparative Risk Assessment of Burden of Disease and Injury Attributable to 67 Risk Factors and Risk Factor Clusters in 21 Regions, 1990–2010: A Systematic Analysis for the Global Burden of Disease Study 2010," *The Lancet* 380, no. 9859 (2012): 2224–2260.

4. H. Frumkin, J. Hess, G. Luber, J. Malilay, and M. McGeehin, "Climate Change: The Public Health Response," *American Journal of Public Health* 98, no. 3 (2008): 435–445.

5. D. Campbell-Lendrum, D. Chadee D, Y. Honda Y, et al., "Human Health: Impacts, Adaptation, and Co-Benefits," in *Climate Change 2014: Impacts, Adaptation, and Vulnerability; Part A: Global and Sectoral Aspects; Contribution of*

*Working Group II to the Fifth Assessment Report of the Intergovernmental Panel on Climate Change*, ed. C. B. Field, V. R. Barros, D. J. Dokken, et al., 709–754 (New York: Cambridge University Press, 2014).

6. Parry, Lowe, and Hanson, "Overshoot, Adapt and Recover," 1102.

7. UNFCCC, "Investment and Financial Flows to Address Climate Change: An Update," United Nations Framework Convention on Climate Change Technical Report FCCC/TP/2008/7, November 26, 2008.

8. S. C. Sherwood and M. Huber, "An Adaptability Limit to Climate Change due to Heat Stress, *Proceedings of the National Academy of Sciences* 107, no. 21 (2010): 9552–9555.

9. J. Schwartz, "Freight Train Late? Blame Chicago," *New York Times*, July 5, 2012.

10. World Health Organization, *World Malaria Report, 2014* (Geneva: World Health Organization, 2014).

11. Ibid.

12. Centers for Disease Control and Prevention, "Insecticide Treated Nets," Centers for Disease Control and Prevention, last modified December 28, 2015, http://www.cdc.gov/malaria/malaria_worldwide/reduction/itn.html.

13. Ibid.

14. World Health Organization, *World Malaria Report, 2014.*

15. M. C. Thomson, F. J. Doblas-Reyes, S. J. Mason, et al., "Malaria Early Warnings Based on Seasonal Climate Forecasts from Multi-Model Ensembles," *Nature* 439, no. 7076 (2006): 576–579.

16. S. T. Agnandji, B. Lell, S. S. Soulanoudjingar, et al., "First Results of Phase 3 Trial of RTS,S/AS01 Malaria Vaccine in African Children," *New England Journal of Medicine* 365, no. 20 (2011): 1863–1875.

17. D. D. Chadee, "Key Premises: A Guide to Aedes Aegypti (Diptera: Culicidae) Surveillance and Control," *Bulletin of Entomological Research* 94, no. 3 (2004): 201–207.

18. World Health Organization, *World Malaria Report, 2013* (Geneva: World Health Organization, 2013).

19. D. A. Focks, R. J. Brenner, J. Hayes, and E. Daniels, "Transmission Thresholds for Dengue in Terms of *Aedes aegypti* Pupae per Person with Discussion of Their Utility in Source Reduction Efforts," *The American Journal of Tropical Medicine and Hygiene* 62, no. 1 (2000): 11–18.

20. Chadee, "Key Premises."

21. D. D. Chadee, B. Shivnauth, S. C. Rawlins, and A. A. Chen, "Climate, Mosquito Indices and the Epidemiology of Dengue Fever in Trinidad (2002–2004)." *Annals of Tropical Medicine and Parasitology* 101, no. 1 (2007): 69–77.

22. D. J. Gubler and G. G. Clark, "Community Involvement in the Control of Aedes Aegypti," *Acta Tropica* 61, no. 2 (1996): 169–179.

23. L. Villar, G. H. Dayan, J. L. Arredondo-García, et al., "Efficacy of a Tetravalent Dengue Vaccine in Children in Latin America," *New England Journal of Medicine* 372, no. 2 (2015): 113–123.

24. M. R. Capeding, N. H. Tran, S. R. Hadinegoro, et al., "Clinical Efficacy and Safety of a Novel Tetravalent Dengue Vaccine in Healthy Children in Asia: A Phase 3, Randomized, Observer-Masked, Placebo-Controlled Trial," *The Lancet* 384, no. 9951 (2014): 1358–1365.

25. J. R. Porter, L. Xie, A. J. C. K. Challinor, et al., "Food Security and Food Production Systems," in *Climate Change 2014: Impacts, Adaptation, and Vulnerability; Part A: Global and Sectoral Aspects; Contribution of Working Group II to the Fifth Assessment Report of the Intergovernmental Panel on Climate Change*, ed. C. B. Field, V. R. Barros, D. J. Dokken, et al., 485–533 (New York: Cambridge University Press, 2014).

26. D. P. Garrity, F. K. Akinnifesi, O. C. Ajayi, et al., "Evergreen Agriculture: A Robust Approach to Sustainable Food Security in Africa," *Food Security* 2, no. 3 (2010): 197–214.

27. Ibid.

28. J. Sendzimir, C. P. Reij, and P. Magnuszewski, "Rebuilding Resilience in the Sahel: Regreening in the Maradi and Zinder Regions of Niger," *Ecology and Society* 16, no. 3 (2011): 1, http://dx.doi.org/10.5751/ES-04198-160301.

29. Garrity, Akinnifesi, Ajayi, et al., "Evergreen Agriculture."

30. J. P. Mulder, S. Hommes, and E. M. Horstman, "Implementation of Coastal Erosion Management in the Netherlands," *Ocean and Coastal Management* 54, no. 12 (2011): 888–897.

31. Ibid.

32. K. de Bruin, R. B. Dellink, A. Ruijs, et al., "Adapting to Climate Change in the Netherlands: An Inventory of Climate Adaptation Options and Ranking of Alternatives," *Climatic Change* 95, nos. 1–2 (2009): 23–45.

33. T. Korten, "In Florida, Officials Ban Term 'Climate Change,'" *Miami Herald*, March 8, 2015.

34. P. Mozumder, E. Flugman, and T. Randhir, "Adaptation Behavior in the Face of Global Climate Change: Survey Responses from Experts and Decision Makers Serving the Florida Keys," *Ocean and Coastal Management* 54, no. 1 (2011): 37–44.

35. Southeast Regional Climate Change Compact Counties, "A Region Responds to a Changing Climate," *Southeast Florida Regional Compact Climate Change*, October 2012, http://www.southeastfloridaclimatecompact.org/wp-content/uploads/2014/09/regional-climate-action-plan-final-ada-compliant.pdf.

36. West Virginia Division of Culture and History, "Buffalo Creek," West Virginia Archives and History, 2010, http://www.wvculture.org/history/buffcreek/bctitle.html.

37. US Environmental Protection Agency, "Carbon Pollution Emission Guidelines for Existing Stationary Sources: Electric Utility Generating Units;

Final Rule," EPA, 2014, http://www.epa.gov/cleanpowerplan/clean-power-plan
-existing-power-plants.

38. M. Z. Jacobson, R. W. Howarth, M. A. Delucchi, et al., "Examining the
Feasibility of Converting New York State's All-Purpose Energy Infrastructure to
One Using Wind, Water, and Sunlight," *Energy Policy* 57 (2013): 585–601.

39. Committee on Geoengineering Climate: Technical Evaluation and Discussion
of Impacts, *Climate Intervention: Reflecting Sunlight to Cool Earth* (Washington,
DC: National Academies Press, 2015); Committee on Geoengineering Climate:
Technical Evaluation and Discussion of Impacts, *Climate Intervention: Carbon
Dioxide Removal and Reliable Sequestration* (Washington, DC: National Acade-
mies Press, 2015).

40. K. H. Coale, K. S. Johnson, S. E. Fitzwater, et al., "A Massive Phytoplankton
Bloom Induced by an Ecosystem-Scale Iron Fertilization Experiment in the Equatorial Pacific Ocean," *Nature* 383 (October 10, 1996): 495–501.

41. D. Aldridge, "A Guide to Iron Fertilisation of the Ocean," *Words in mOcean*,
October    18,    2012,    http://wordsinmocean.com/2012/10/18/a-guide-to-iron
-fertilisation-of-the-ocean.

42. L. J. Smith and M. S. Torn, "Ecological Limits to Terrestrial Biological Carbon
Dioxide Removal," *Climatic Change* 118, no. 1 (2013): 89–103.

43. A. H. Lockwood, *The Silent Epidemic: Coal and the Hidden Threat to Health*
(Cambridge, MA: MIT Press, 2012).

44. N. J. Langford, "Carbon Dioxide Poisoning," *Toxicological Reviews* 24,
no. 4 (2005): 229–235.

45. P. J. Baxter, M. Kapila, and D. Mfonfu, "Lake Nyos Disaster, Cameroon,
1986: The Medical Effects of Large Scale Emission of Carbon Dioxide?," *The
BMJ* 298, no. 6685 (1989): 1437–1441.

46. United States Geological Survey, "Oklahoma Earthquake Information," last
updated April 18, 2014, http://earthquake.usgs.gov/earthquakes/states/?region
=Oklahoma, accessed December 29, 2015.

47. J. B. Stewart, "King Coal, Long Besieged, Is Deposed by Market," *New York
Times*, August 6, 2015.

48. P. Minnis, E. F. Harrison, L. L. Stowe, et al., "Radiative Climate Forcing by
the Mount Pinatubo Eruption," *Science* 259, no. 5100 (1993): 1411–1415.

49. Committee on Geoengineering Climate: Technical Evaluation and Discussion
of Impacts, *Climate Intervention: Reflecting Sunlight to Cool Earth*; Committee
on Geoengineering Climate: Technical Evaluation and Discussion of Impacts,
*Climate Intervention*.

50. Jacobson Howarth, Delucchi, et al., "Examining the Feasibility of Converting
New York State's All-Purpose Energy Infrastructure."

51. K. Korosec, "Clues Emerge for Tesla's $5 Billion Battery Factory," *Fortune*,
March 10, 2014.

52. H. Fountain, "Liquid Batteries for Solar and Wind Power," *New York Times*,
April 22, 2015.

53. J. Liu, Y. Liu, N. Liu, et al., "Metal-Free Efficient Photocatalyst for Stable Visible Water Splitting via a Two-Electron Pathway," *Science* 347, no. 6225 (2015): 970–974.

54. N. Gingrich, "Newt Gingrich: Double the N.I.H. Budget," *New York Times*, April 22, 2015.

55. The complete text can be found at https://www.whitehouse.gov/the-press -office/2011/01/25/remarks-president-state-union-address.

56. Potsdam Institute for Climate Impact Research and Climate Analytics, *Turn Down the Heat: Why a 4° C Warmer World Must be Avoided* (Washington, DC: International Bank for Reconstruction and Development/The World Bank, 2012).

57. C. C. Jaeger and J. Jaeger, "Three Views of Two Degrees," *Regional Environmental Change* 11, no. 1 (2011): 15–26.

58. T. F. Stocker, D. Qin, G-K. Plattner, et al., eds., *Climate Change 2013: The Physical Science Basis; Contribution of Working Group I to the Fifth Assessment Report of the Intergovernmental Panel on Climate Change*, 70 (New York: Cambridge University Press, 2014).

# Index

Printed in the United States
by Baker & Taylor Publisher Services